Nanoscale Standards by Metrological AFM and Other Instruments

Ichiko Misumi

IOP | ebooks

Nanoscale Standards by Metrological AFM and Other Instruments

Nanoscale Standards by Metrological AFM and Other Instruments

Ichiko Misumi
National Institute of Advanced Industrial Science and Technology, Tsukuba, Ibaraki, Japan

IOP Publishing, Bristol, UK

ISBN 978-0-7503-3191-3 (ebook)
ISBN 978-0-7503-3189-0 (print)
ISBN 978-0-7503-3192-0 (myPrint)
ISBN 978-0-7503-3190-6 (mobi)

DOI 10.1088/978-0-7503-3191-3

Version: 20210501

IOP ebooks

British Library Cataloguing-in-Publication Data: A catalogue record for this book is available from the British Library.

Published by IOP Publishing, wholly owned by The Institute of Physics, London

IOP Publishing, Temple Circus, Temple Way, Bristol, BS1 6HG, UK

US Office: IOP Publishing, Inc., 190 North Independence Mall West, Suite 601, Philadelphia, PA 19106, USA

Contents

Preface

Recently, there have been advances in nanometrology, and increased demand for the reliability of measurement results. In order to ensure the reliability of measurement results, it is necessary to calibrate the measurement apparatus using a nano dimensional standard. To use nano dimensional standards, it is desirable to understand how standards are calibrated. The author has developed metrological atomic force microscopes (metrological AFMs) and has been involved in the calibration of various nano dimensional standards using metrological AFMs. This book describes the current status and future trends of various nano dimensional standards. This book will be of interest to users of nano dimensional standards, particularly those concerned about how to use nano dimensional standards and how to calibrate a nano dimensional standard developed by the user. Readers of this book will be able to deepen their understanding of nano dimensional standards, which will lead to improvements in quality inspection and yield of the product.

The contents of this book are as follows. (1) Nano dimensional standards: their importance is explained. Metrological AFM, one of the calibration instruments, is introduced. Furthermore, the fundamental reference document, the Guide to the Expression of Uncertainty in Measurement (ISO/IEC Guide 98-3:2008) is explained briefly. (2) Pitch of one-dimensional grating: one-dimensional grating is one of the most important nano dimensional standards. Pitch calibration methods by a metrological AFM and optical diffractometer are explained. (3) Step height: Step height is also one of the most important nano dimensional standards. The calibration method by a metrological AFM based on ISO 5436-1 is explained. Atomic step is also introduced. (4) Pitch of two-dimensional grating: pitch calibration of a two-dimensional grating has a reliability issue that is different from that of a one-dimensional grating since scanning of a probe would become unstable unless the scanning direction is perfectly perpendicular to the sidewalls of the two-dimensional grating patterns. The pitch calibration method of two-dimensional grating by a metrological AFM is explained. (5) Line width: line width is greatly affected by the AFM tip shape. Special metrological AFMs (CD-AFMs and tilted-type AFMs) are explained. (6) Nano particle size: nano particle size has been calibrated by other types of instruments, for instance, dynamic light scattering (DLS) and differential mobility analysis (DMA). Recently, a metrological AFM has also been used for nano particle size measurements. (7) Surface roughness: surface roughness has been measured by other types of instruments, for instance, stylus and optical instruments. Recently, a metrological AFM has also been used for surface roughness measurements. (8) Secondary realization of the SI meter using silicon lattice parameters is explained. Furthermore, the demands of industry and future directions are described.

Acknowledgments

Part of the author's research activity is supported by the New Energy and Industrial Technology Development Organization (NEDO) (project codes P02045, 00X50023x), Nippon Sheet Glass Foundation for Materials Science and Engineering, and The Precise Measurement Technology Promotion Foundation. The author is highly appreciative of discussions with Azuma Y, Ehara K, Fujimoto H, Fujimoto T, Gonda S, Hirai A, Ichimura S, Itoh H, Kinoshita K, Kizu R, Kojima I, Kurosawa T, Lu M, Naoi K, Sato O, Sugawara K, Takahashi K, Takahata K, Takatsuji T, Tanaka H, Tanimura Y and Yoshizaki K, all from the AIST, Japan, as well as Takamasu K and Takahashi S from the University of Tokyo, Japan, Kitta J (JQA, Japan), Nakayama Y (Hitachi, Ltd., Japan), Kawada H (Hitachi High-Tech Corporation, Japan), Kurihara K, Kurosawa S, Takano A and Takenaka H from NTT-AT, Japan, Ito Y and Omote K from Rigaku Corporation, Japan, Bosse H, Dai G, Danzebrink H-U, Dziomba T and Koenders L from PTB, Germany, Dagata J, Dixson R and Orji N from NIST, US, Fu W E (CMS, ITRI, Taiwan), Yacoot A (NPL, UK), Garnaes J (DFM, Denmark), Jin J and Park J from KRISS, Korea, and Langham C S (Nihon University, Japan). The author is grateful to Navas J (IOP Publishing, UK) for the opportunity to write this book.

Author biography

Ichiko Misumi

Dr Misumi graduated with a degree in precision machinery engineering, from the University of Tokyo in 1996 and obtained her Doctor of Engineering from the University of Tokyo in 2004. She joined the National Research Laboratory of Metrology (NRLM) in 2000. In 2001, NRLM was renamed the National Metrology Institute of Japan (NMIJ). She received the 'Outstanding Precision Measurement Paper Award 2005, Measurement Science and Technology, Institute of Physics (IOP)', and 'Highly Commended Paper Award 2007, Measurement Science and Technology, Institute of Physics (IOP)'. She also received the 'Award for Science and Technology (Development Category), The Commendation for Science and Technology by the Minister of Education, Culture, Sports, Science and Technology' in 2011.

IOP Publishing

Nanoscale Standards by Metrological AFM and Other Instruments

Ichiko Misumi

Chapter 1

Nano dimensional standards

Examples of nano dimensional standards are shown. Metrological atomic force microscopes (metrological AFMs), one of the calibration instruments for nano dimensional standards, are explained. Furthermore, the fundamental reference document, 'The Guide to the Expression of Uncertainty in Measurement (ISO/IEC Guide 98-3:2008)' is introduced briefly.

1.1 Nano dimensional standards

Nanotechnology has been advancing rapidly. Several types of nanostructures have been developed and used as devices. The size of a nanostructure is measured as a means for confirming whether the nanostructure has the functions as designed. That is, the size of the nanostructure may directly affect its performance. Therefore, it can be said that nanometrology has also become important.

Several types of nanometrological instruments are used in nanostructure manufacturing sites. To ensure the reliability of nanometrology, it is necessary to calibrate nanometrological instruments with nano dimensional standards. Nano dimensional standards are nano dimensional specimens which have been calibrated by SI (the International System of Units) traceable nanometrological instruments.

Figure 1.1 shows examples of nano dimensional standards. A schematic drawing of a one-dimensional (1D) grating standard is shown in (a). The pitch of 1D grating is defined as the distance between the centers of the neighboring line patterns. A schematic drawing of a step height standard is shown in (b). Firstly, the base line is determined by the left and the right areas of a line pattern. Step height is defined as the distance from the base line to the height of the center of the line pattern. (a) 1D grating standard, and (b) step height standard are the most basic nano dimensional standards. Users can calibrate their nanometrological instruments in x-, y-, and z-axes directions using a 1D grating standard and a step height standard. Here, it is necessary

doi:10.1088/978-0-7503-3191-3ch1

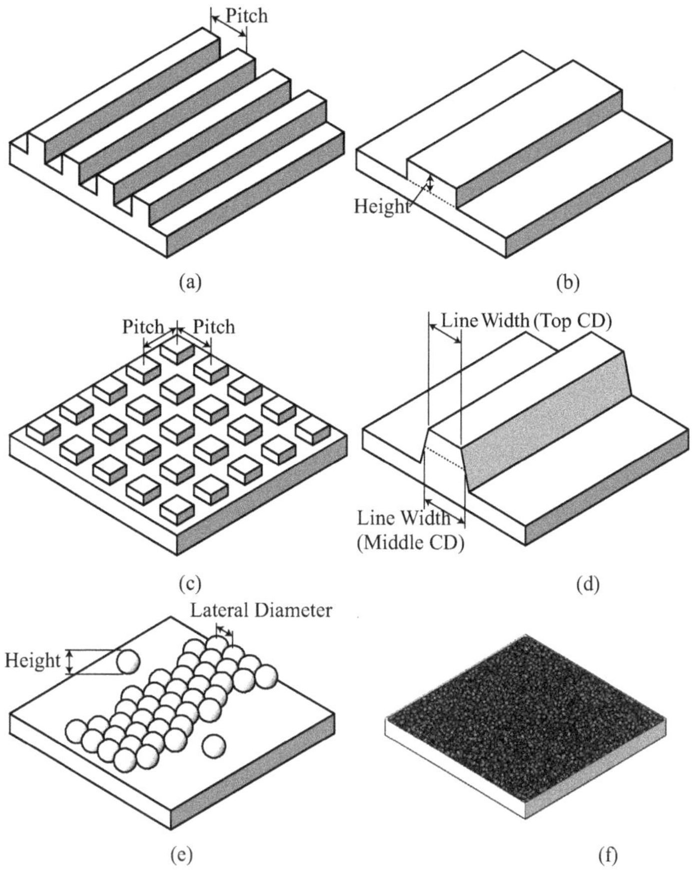

Figure 1.1. Examples of nano dimensional standards. (a) A schematic drawing of a one-dimensional (1D) grating standard. The pitch of 1D grating is defined as the distance between the centers of the neighboring line patterns. (b) A schematic drawing of a step height standard. Firstly, the base line is determined by the left and the right areas of a line pattern. Step height is defined as the distance from the base line to the height of the center of the line pattern. (c) A schematic drawing of a two-dimensional (2D) grating standard. The pitch of 2D grating is defined as the distance between the centers of the convex parts in the x- or y-direction. (d) A schematic drawing of a line width standard. There are several definitions of line width. Top critical dimension (CD) or line width is defined as the distance between the top edges of a line pattern. It is determined that the distance between the top surface of a line pattern and a base plane consisting of left and right base areas of a line pattern is h. The middle CD or line width is defined as the distance between two sidewalls at the height of $h/2$. (c) A schematic drawing of standard nano particles. Suspension with standard nano particles is put on a flat substrate, for instance mica, and the specimen is dried. The height of a particle is defined as the distance between the top of a particle and the substrate surface. Nano particles sometimes make a close-packed structure. Lateral diameter is defined as the distance between the centers of neighboring particles. (f) Schematic drawing of surface roughness standard.

to re-install a 1D grating standard in the x- or the y-axis direction of a nanometrological instrument. When users have two 1D grating standards, they install them so that the setting direction of a grating is perpendicular to the other one and they can calibrate the x- and the y-axes of their nanometrological instruments simultaneously.

A schematic drawing of a two-dimensional (2D) grating standard is shown in (c). The pitch of 2D grating is defined as the distance between the centers of the convex parts in the *x*- or *y*-direction. Users can use a 2D grating standard instead of two 1D grating standards. A schematic drawing of a line width standard is shown in (d). There are several definitions of line width. Top critical dimension (CD) or line width is defined as the distance between the top edges of a line pattern. It is determined that the distance between the top surface of a line pattern and a base plane consisting of left and right base areas of a line pattern is *h*. The middle CD or line width is defined as the distance between two sidewalls at the height of $h/2$. A schematic drawing of standard nano particles is shown in (e). Suspension with standard nano particles is placed on a flat substrate, for instance mica, and the specimen is dried. The height of a particle is defined as the distance between the top of a particle and the substrate surface. Nano particles sometimes make a close-packed structure. Lateral diameter is defined as the distance between the centers of neighboring particles. A schematic drawing of surface roughness standard is shown in (f). Most surface roughness standards are manufactured for stylus-type instruments. However, surface roughness standards for optical instruments and AFMs have also been developed recently. Precise explanations of these nano dimensional standards are given in chapters 2, 3, 4, 5, 6, and 7.

Ordinally, a calibration certificate of a nano dimensional standard shows a calibrated value and its uncertainty, which can be used as they are. However, it is recommended that users understand how to calibrate nano dimensional standards. Users are sometimes confused when they see different calibrated values by several types of nanometrological instruments in a certificate of a nano dimensional standard. A nano particle size comparison measurement (APMP.L-S5) performed between national metrology institutes (NMIs) is shown as an example (Lin *et al* 2019). In this comparison, nano particles of three different materials (Au 10 nm, Ag 20 nm, polystyrene latex (PSL) 30 nm, PSL 100 nm, PSL 300 nm) with sizes ranging from 10 nm to 300 nm were used. Selected nano particles were measured by AFM, electron microscope (EM), differential mobility analyzer (DMA), small angle X-ray scattering (SAXS) and dynamic light scattering (DLS). Figure 1.2 shows the method-dependent reference value (MRV) of PSL 100 nm. The MRV was determined according to the measurement results and uncertainties were reported by the participants. Each measurement method has its own MRV. It can be seen that the measured values differ depending on the measuring method.

1.2 Metrological atomic force microscopes

Metrological atomic force microscopes (metrological AFMs) are typical devices used for the calibration of nano dimensional standards. They differ from ordinary AFMs in that the displacement of scanner(s) or probe(s) is measured by laser interferometer(s) precisely. Measurement results obtained by metrological AFMs are SI traceable. The advantage of metrological AFMs is that several types of nanometrological standards can be calibrated. Furthermore, conductive materials and transparent materials can be measured by metrological AFMs. The disadvantage is the low throughput.

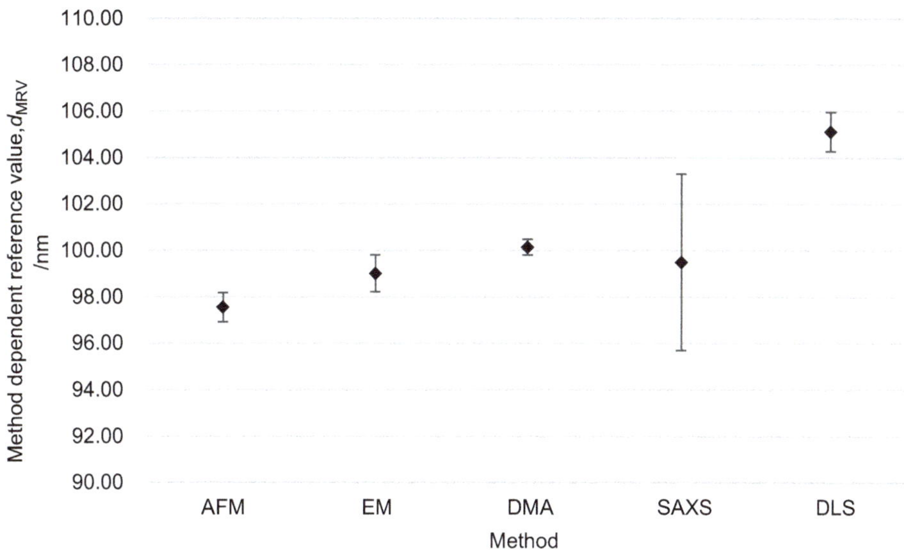

Figure 1.2. Method-dependent reference value d_{MRV} and combined standard uncertainty u_c.

In addition, it is often not compatible with wafer-sized samples. Metrological AFMs have become reference standards for other nanometrological instruments.

Yacoot and Koenders (2011) reviewed metrological scanning tunneling microscopes (metrological STMs) and metrological AFMs from the dawn of metrological scanning probe microscopes to around 2011. NIST, NMI in the USA, developed a molecular measuring machine (M^3) for the first time as a metrological scanning probe microscope in 1989 (Teague 1989, Kramar *et al* 2011). The M^3 is an ultra-high precision plane coordinate measuring machine that can position and measure with atomic scale accuracy in the 2500 mm^2 area. A probe based on either a scanning tunneling microscope, or an AFM, was incorporated into a very stable core mechanical structure. The M^3 realized both wide scanning range and atomic-resolution metrology (Kramar *et al* 2011).

After that, NIST developed an AFM capable of accurate dimensional measurements (Schneir *et al* 1994, Kramar *et al* 2011). This system is called 'calibrated AFM (C-AFM)'. C-AFM was built as far as possible from off-the-shelf components. It uses a flexure stage driven by a piezoelectric transducer for scanning, a heterodyne interferometer to measure the *X* and *Y* displacements of the sample, a capacitive sensor to measure the *Z* displacement of the sample, and a commercial AFM control system. The control system has two feedback loops that read from the *X* and *Y* interferometers, respectively, and adjust the piezo voltage to keep the *X*–*Y* scan position accurate.

PTB, NMI in Germany, developed a metrological STM in 1994 (Jusko *et al* 1994). In addition, PTB realized the 3D calibration of the scanning force microscope (SFM) by incorporating the laser interferometric 3D measurement system in the SFM and by compensating for the calibration error measured by implementing the function or data corresponding to the control software of the SFM (Hasche *et al* 2002). By using the metrological SFM, it became possible to measure dimensions with an uncertainty of a few nanometers. The metrological SFM was used to determine the surface morphology of some objects such as microhard indenters, optical gratings, and film thicknesses. It was confirmed that the results agreed by performing comparison measurements of the step heights using a metrological SFM and an interference optical microscope.

METAS (former OFMET), NMI in Switzerland, developed a long-range AFM profiler system based on a commercially available AFM and a homemade linear sample displacement stage (Meli and Thalmann 1998). The AFM head contains a parallelogram type scanner with capacitive position sensors on all three axes. The reference cube, near the tip, acts as a counter electrode for the capacitive sensor. Below this AFM head, a linear sample displacement stage consisting of a monolithic flexure forming a double parallelogram was placed. This piezo-driven stage offers highly linear motion in excess of 380 μm. Its displacement is measured simultaneously by a capacitive position sensor and a differential double-pass plane mirror interferometer. Both measurement systems have sub-nanometer resolution. Two modes of operation are possible for measuring periodic structures. There is a direct scan mode, in which the position of the displacement stage increases point by point, while the AFM head measures the height, and a compound scan mode, in which the displacement stage produces an offset. Pitch comparison measurements of a one-dimensional grating using a long-range AFM profile system and a laser diffractometer showed extremely good agreement.

NMIJ, AIST (formerly NRLM, AIST), developed an AFM equipped with a high-resolution three-axis laser interferometer (Gonda *et al* 1999). Figures 1.3 and 1.4 show the cross-section and block diagram of the nanometrological AFM, respectively (Misumi *et al* 2003). The nanometrological AFM system is a stage scan type and operates in contact AFM mode and DC mode. The nanometrological AFM system consists of a stage unit, an AFM probe unit, and an interferometer unit. The stage unit consists of a piezo-driven XY axes leaf spring stage and a Z-axis tube piezo actuator. The scanning range of this stage unit is approximately 17.5 μm (X) × 17.5 μm (Y) × 2.5 μm (Z). The bending mirror of the Z-axis interferometer is integrated into the Z-axis scanner tube. A three-sided moving mirror for the XYZ interferometer unit is installed on top of the Z-axis scanner. There are four optical paths in each axis of the interferometer, and the theoretical resolution of the interferometer unit is approximately 40 pm. The sample is mounted inside a three-sided mirror and mounted on an ultrasonically driven rotary stage that adjusts the lateral rotation angle of the sample. Atomic forces are detected using the standard optical lever method. The laser source for the interferometer unit is a practical frequency-stabilized He–Ne laser with a wavelength of 633 nm (Model 117A, Spectra-Physics Ltd). Laser frequencies are calibrated using an I_2 stabilized He–Ne laser before measuring nano dimensional standards. The position of the stage is servo-controlled using the interferometer signal in real time and the metrological AFM can meet nanometer standards for direct length standard traceable calibration.

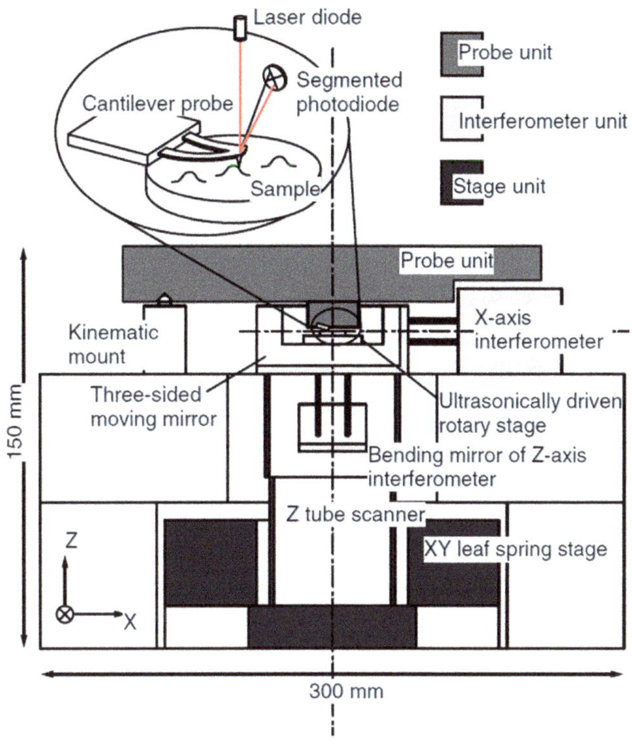

Figure 1.3. Cross-sectional drawing of the main unit of the nanometrological AFM. It consists of a stage unit, an AFM probe unit, and interferometer units (Misumi *et al* 2003). Copyright IOP Publishing Ltd. All rights reserved.

1.3 The Guide to the Expression of Uncertainty in Measurement (The GUM, ISO/IEC Guide 98-3:2008)

Since chapter 2 of this book, there have been many explanations about the uncertainty evaluation of nano dimensional standard measurements using metrological AFMs. The uncertainty evaluations are based on the GUM (The Guide to the Expression of Uncertainty in Measurement (ISO/IEC Guide 98-3:2008)). The GUM is the principle document in uncertainty evaluation. If users are not familiar with uncertainty evaluation, they should refer to the GUM.

1.4 Summary

Six types of nano dimensional standards were introduced as examples. Next, metrological AFMs, calibration instruments for nano dimensional standards, were explained. Furthermore, the fundamental reference document, 'The Guide to the Expression of Uncertainty in Measurement (ISO/IEC Guide 98-3:2008)' was explained briefly.

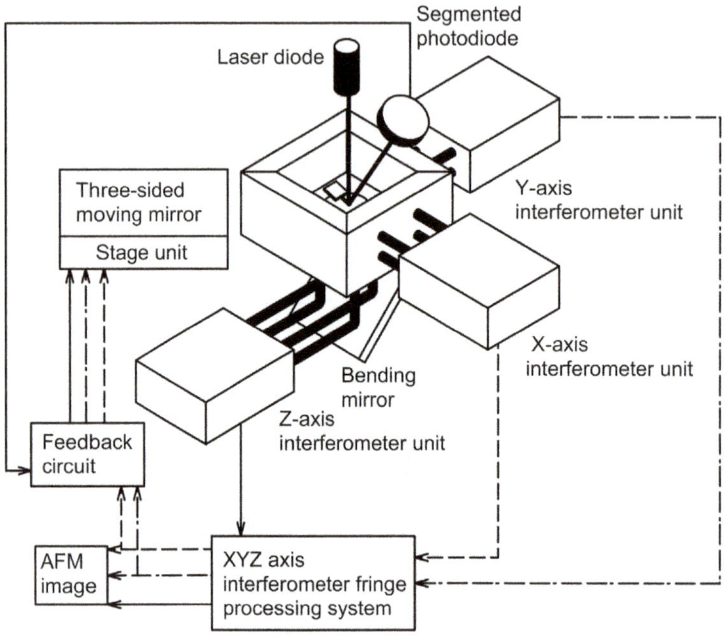

Figure 1.4. Block diagram of the nanometrological AFM system. The stage position is servo-controlled using the interferometer signals in real time (Misumi *et al* 2003). Copyright IOP Publishing Ltd. All rights reserved.

References

Gonda S, Doi T, Kurosawa T, Tanimura Y, Hisata N, Yamagishi T, Fujimoto H and Yukawa H 1999 Real-time, interferometrically measuring atomic force microscope for direct calibration of standards *Rev. Sci. Instrum.* **70** 3362–8

Hasche K, Herrmann K, Mirandé W, Seemann R, Vitushkin L, Xu M and Yu G 2002 Calibrated scanning force microscope with capabilities in the sub-nanometre range *Surf. Interface Anal.* **33** 71–4

ISO/IEC GUIDE 98-3 2008 Uncertainty of Measurement—Part 3: Guide to the expression of uncertainty in measurement (GUM:1995)

Jusko O, Zhao X, Wolff H and Wilkening G 1994 Design and three dimensional calibration of a measuring scanning tunneling microscope for metrological applications *Rev. Sci. Instrum.* **65** 2514–8

Kramar J A, Dixson R and Orji N G 2011 Scanning probe microscope dimensional metrology at NIST *Meas. Sci. Technol.* **22** 024001

Lin H-L *et al* 2019 Nanoparticle characterization—supplementary comparison on nanoparticle size *Metrologia* **56** 1A

Meli F and Thalmann R 1998 Long range AFM profiler used for accurate pitch measurements *Meas. Sci. Technol.* **9** 1087–92

Misumi I, Gonda S, Kurosawa T and Takamasu K 2003 Uncertainty in pitch measurements of one-dimensional grating standards using a nanometrological atomic force microscope *Meas. Sci. Technol.* **14** 463–71

Schneir J, McWaid T H, Alexander J and Wilfley B P 1994 Design of an atomic force microscope with interferometric position control *J. Vac. Sci. Technol. B* **12** 3561–6

Teague E C 1989 The National Institute of Standards and Technology molecular measuring machine project: metrology and precision engineering design *J. Vac. Sci. Technol.* B **7** 1898–902

Yacoot A and Koenders L 2011 Recent developments in dimensional nanometrology using AFMs *Meas. Sci. Technol.* **22** 122001

Ichiko Misumi

Chapter 2

One-dimensional grating

One-dimensional grating is one of the most important nano dimensional standards. Pitch calibration methods by a metrological atomic force microscope and optical diffractometer are explained.

2.1 Importance of one-dimensional grating as a nano dimensional standard

One-dimensional grating is one of the most important nano dimensional standards. Critical dimension scanning electron microscopes (CD-SEMs) are often used as inspection instruments at semiconductor device manufacturing sites. The CD-SEM is a high-throughput and highly repetitive instrument, but it requires periodic magnification calibration. One-dimensional grating is often used for CD-SEM magnification calibration (Nakayama *et al* 1988). Figure 2.1 shows an example of a one-dimensional grating for magnification calibration of CD-SEM (Misumi *et al* 2003b). The one-dimensional grating itself should be manufactured with high precision and the pitch precisely calibrated. First, we will introduce a method of making a one-dimensional grating.

2.1.1 Laser interferometer lithography

A submicron-level reference for electron beam metrology systems was proposed in 1988 by Nakayama and Okazaki. This reference is a fine rectangular profile grating made by laser interferometer lithography and anisotropic chemical etching of (110) crystalline silicon. The pitch size of the grating is derived from the wavelength of the laser and the angle of incidence of holographic lithography. The system is adjusted so that the pitch is approximately 240 nm. The deep lamellar lattice created by anisotropic chemical etching of (110) Si produces stable, high-contrast secondary electron signals in electron beam metrology systems. Optical diffractometry ensures

doi:10.1088/978-0-7503-3191-3ch2

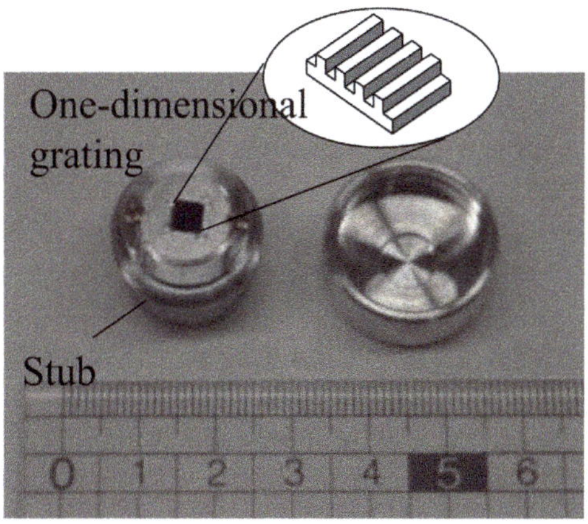

Figure 2.1. Example of a one-dimensional grating for magnification calibration of CD-SEM (Misumi *et al* 2003b). Copyright IOP Publishing Ltd. All rights reserved.

the correct grating pitch size. The measurement error is estimated to be less than 1 nm. Reference pitch size measurement with optical diffraction and electron beam metrology systems guarantees a pitch size error of less than 20 nm.

2.1.2 Laser-focused atomic deposition

Laser-focused atomic deposition was proposed by McClelland *et al* in 1993 as a method for producing stable nanometer-sized structures in air. Laser light can be used to control the movement of atoms at the nanoscale. Chromium atoms were focused by a standing wave laser field as they were deposited on a silicon substrate. Atoms are collimated in a lateral laser-cooled region before being focused by a standing wave. The frequency of the dye laser is 60–240 MHz higher than the atomic resonance. The laser-cooled beam is frequency shifted using an acousto-optic modulator (AOM) just below the atomic resonance. The laser-cooled beams are linearly polarized perpendicular to each other by a quarter-wave retarder ($\lambda/4$). The second quarter-wave retarder circularly polarizes the standing wave beam. The resulting nanostructure consisted of a series of fine lines covering 0.4 mm × 1 mm. Atomic force microscope (AFM) measurements showed a line width of 65 ± 6 nm, a spacing of 212.78 nm, and a height of 34 ± 10 nm (McClelland *et al* 1993, 2003).

2.1.3 Electron beam cell projection lithography

Laser interferometer lithography and laser-focused atomic deposition are advantageous in that nanostructures can be formed in a millimeter area. On the other hand, the spatial wavelength of the nanostructure is limited by the laser wavelength. A method that can produce nanostructures with a pitch of 100 nm or less is needed. In 2005, Nakayama realized a 100 nm pitch size silicon grating pattern for the calibration standard of an electron beam (EB) measurement system (Nakayama *et al* 2005). Lattice patterns in the *x*- and *y*-directions with pitch sizes of 240, 200, 160, and 100 nm were written by an EB cell projection writing system using a lattice stencil mask and these resist patterns were transferred by dry etching to a silicon substrate. Pitch size error in the x- and y-directions was within 1 nm and the 1.8 × 1.8 mm chip uniformity was within 3σ and less than 3 nm.

One-dimensional gratings with pitches of 50–100 nm were developed using EB cell projection lithography (Misumi *et al* 2006). Figure 2.2 shows a schematic diagram of a one-dimensional grating. The grating was fabricated on a silicon substrate ((110) plane, size: 10 mm × 10 mm × 0.525 mm) and consisted of a scale pattern, reference pattern, and guide pattern. The pattern was fabricated by EB cell projection lithography and anisotropic etching. The sidewalls of the scale and reference patterns were designed to have (1 1 1) faces. The scale pattern was placed in the center of the substrate, whose area was about 200 μm × 200 μm, and a matrix of 25 × 25 small cells (8 μm × 8 μm) was constructed. Each cell had a grid pattern area (7.2 μm × 7.2 μm) and gaps. Each 7.2 μm × 7.2 μm size scale pattern area was fabricated by a single exposure of electron projection through a scale pattern mask cell. The size of the grid pattern was optimized to eliminate line shape distortion. When the grating pitch was a common denominator of 8 μm and 7.2 μm, e.g. 100 or 50 nm, the position of the gap in all cells was the same. The reference pattern was approximately 1 mm above the scale pattern, and its area was approximately 200 μm × 200 μm. The reference pattern was used to optimize the measurement conditions before measuring the scale pattern. The guide pattern is positioned to allow easy placement of the AFM cantilever probe on the scale or reference pattern. The step height and line width of the scale and reference patterns were approximately 100 nm and 20 nm, respectively. The one-dimensional grating sample was manufactured by Nippon Telegraph and Telephone Advanced Technology (NTT AT) based on the design. Figures 2.3 and 2.4 show

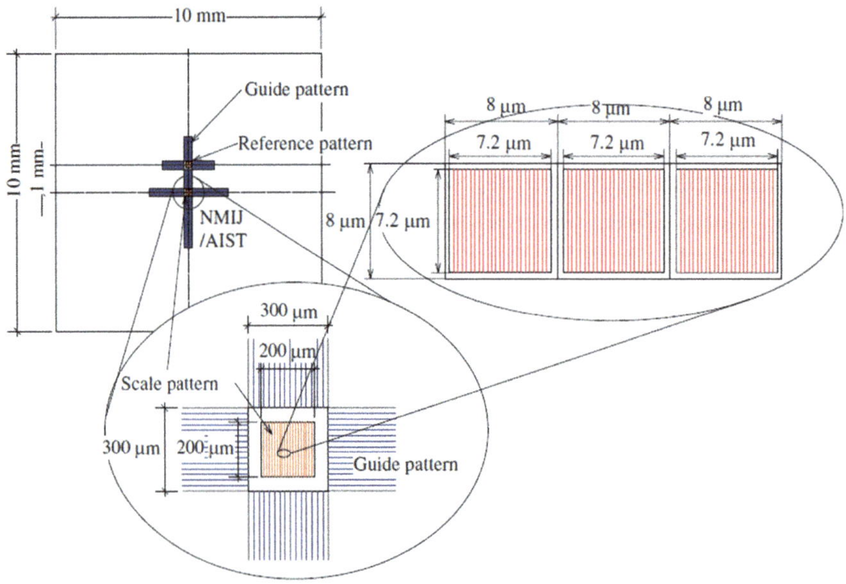

Figure 2.2. Schematic drawing of the one-dimensional grating sample (Misumi *et al* 2006). Copyright IOP Publishing Ltd. All rights reserved.

(*a*) 100 nm pitch (*b*) 50 nm pitch

Figure 2.3. SEM images of the scale pattern area of the samples: (a) 100 nm pitch, (b) 50 nm pitch (Misumi *et al* 2006). Copyright IOP Publishing Ltd. All rights reserved.

(*b*) 50 nm pitch

Figure 2.4. TEM images of the scale pattern area of the samples: (a) 100 nm pitch, (b) 50 nm pitch (Misumi *et al* 2006). Copyright IOP Publishing Ltd. All rights reserved.

scanning electron microscope (SEM) and transmission electron microscope (TEM) images of the manufactured one-dimensional grating, respectively.

2.1.4 One-dimensional grating consisting of a multilayer structure

The minimum pitch that can be produced by EB cell projection lithography is approximately 50 nm (Misumi *et al* 2006). With the progress of nanotechnology, one-dimensional gratings with smaller pitch are required. Some fabrications of nanostructures consisting of multilayer structures have been reported (Senoner *et al* 2005, Itoh *et al* 2006, Misumi *et al* 2007a, 2008). Figure 2.5 shows the fabrication procedure of a one-dimensional grating consisting of a multilayer structure (Misumi *et al* 2008). First, a multilayer thin film made of two types of materials is produced. Next, the wafers are bonded. This is because they are difficult to scan with an AFM when the scale part is at the edge of the sample. The bonded wafer is cut into pieces and the sidewalls of the pieces are polished. Next, the bonded pieces are mounted on the base so that the sample can be easily handled. A periodic structure is obtained by polishing and performing selective etching. Figure 2.6 is a cross-sectional TEM image of a one-dimensional grating consisting of a multilayer film structure. The nominal pitch is 25 nm. The material is a combination of Si and SiO_2. The fabricated one-dimensional grating was used in the bilateral comparison by metrological AFMs between NMIJ and PTB (Misumi *et al* 2010).

2.2 Pitch calibration of one-dimensional grating by metrological AFMs

2.2.1 The center of gravity method

In the secondary electron profile of SEM, for instance, the left edges of the nanostructure are used as the peak positions, and the distance between the adjacent peak positions is calculated as the pitch. On the other hand, in the pitch calculation

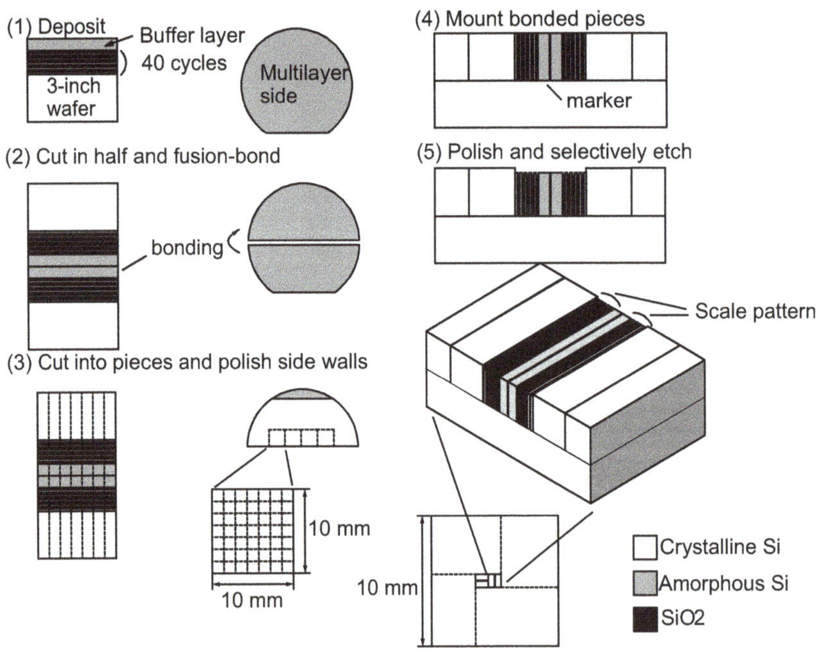

Figure 2.5. Fabrication procedure of a one-dimensional grating consisting of a multilayer structure (Misumi *et al* 2008).

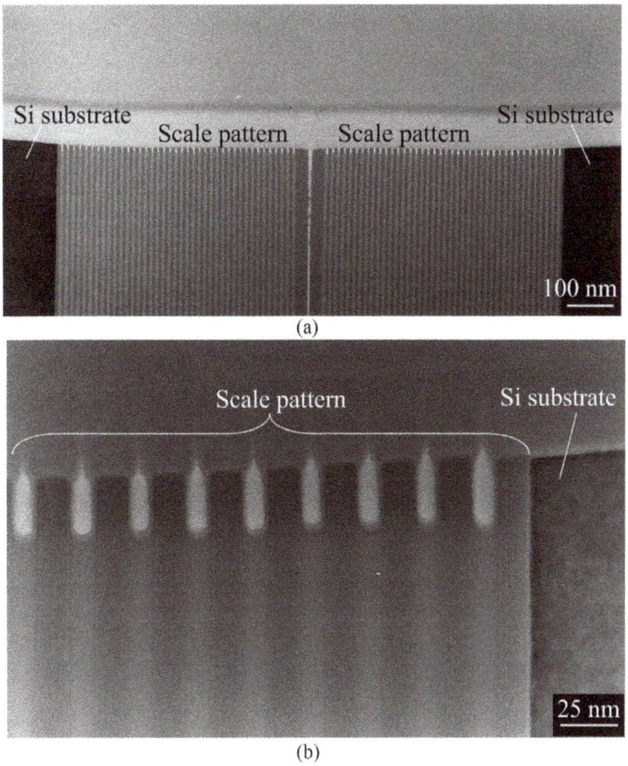

Figure 2.6. Cross-sectional TEM image of a one-dimensional grating consisting of a multilayer film structure. (a) Whole area, and (b) a part of a one-dimensional grating and a silicon substrate (Misumi *et al* 2008). Copyright IOP Publishing Ltd. All rights reserved.

method in the pitch measurement of the one-dimensional grating using a metrological AFM, the method of using the edge positions of the nanostructure as the peak positions is not often applied. A normal AFM controls its system in the Z-axis direction. AFM can stably measure the Z-direction such as step heights and roughness. On the other hand, the AFM control of the sidewall of the nanostructure becomes unstable, and a stable profile cannot be obtained at the sidewall. As a pitch calculation method suitable for an AFM, the center of gravity method is worthy of note. The center of gravity method is as follows. The center of gravity is calculated from the area of a region including the local maximum value of the profile above a certain threshold of the line profile, and the distance between the neighboring centers of gravity is determined as the pitch. METAS (former OFMET) proposed the center of gravity method (Meli and Thalmann 1998). NMIJ (former NRLM), submitted a report on the CCL -S1 (NANO 4) (Meli 2000), the supplementary inter-comparison in the Consultative Committee for Length, using the center of gravity method. NMIJ has described in detail the uncertainty evaluation method and its results for pitch calibration using the center of gravity method (Misumi *et al* 2003a, 2003b). The PTB also uses the center of gravity method (Dai *et al* 2005), and ISO 11952: 2019 also shows the center of gravity method as a method for calculating pitch.

Figure 2.7 shows the procedure for calculating the pitch value (Misumi *et al* 2003a). (1) Figure 2.7(a) shows the line profile obtained by scanning. The contour ridge centerline is obtained by least-squares fitting using peak and valley points. The slope of the line profile is corrected by this centerline. Figure 2.7(b) shows the line profile with tilt correction. The ribs are not exactly parallel to the *Y*-axis. To make the rib parallel to the *Y*-axis direction and to obtain the pitch value, it is necessary to correct the rotation angle in the *XY* plane. (2) Figure 2.7(c) shows the enlarged profile of figure 2.7(b). The area surrounded by the profile curve with local peak

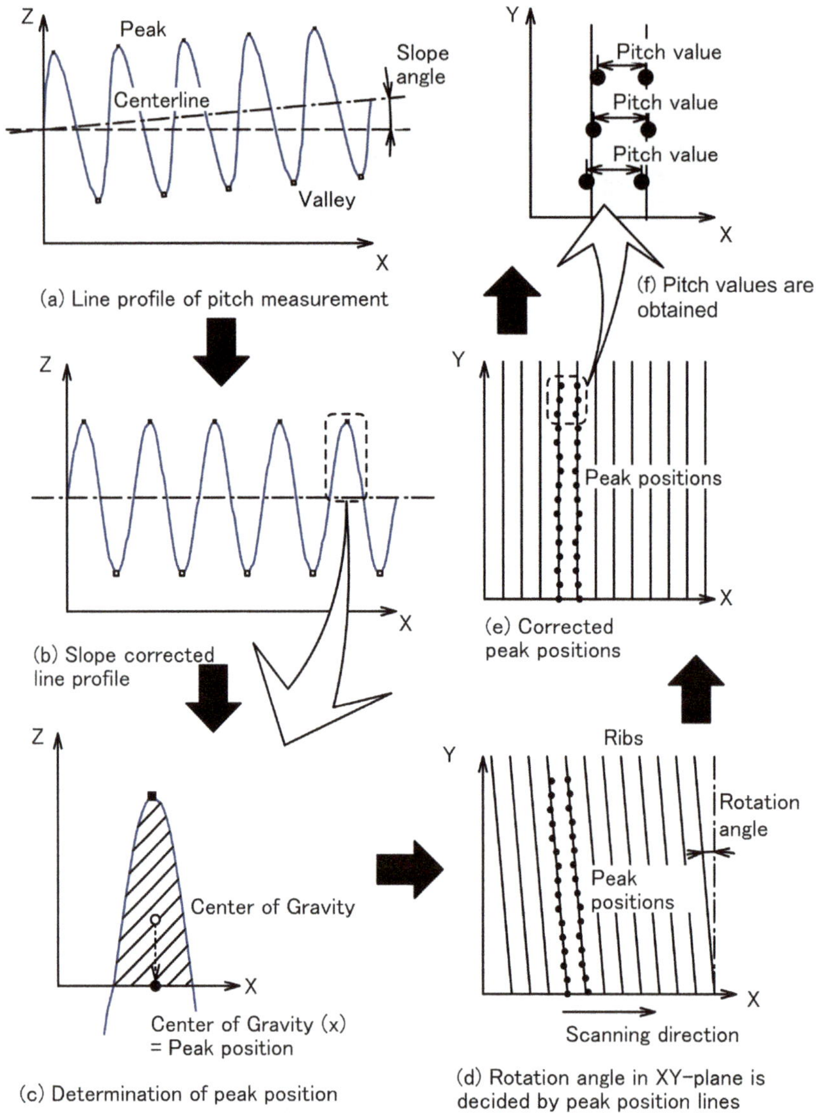

Figure 2.7. Calculation procedure based on the center of gravity method for the pitch values (Misumi *et al* 2003a).

points and baseline (X-axis) is calculated, and the X position of the center of gravity of this area is defined as the representative value of the peak position of each pitch. (3) Figure 2.7(d) shows the peak position line on the XY plane. The scan direction (X-axis direction) and the grid pattern ribs (Y-axis direction) are not exactly perpendicular. Plotting the XY coordinates of the peak positions of all scan lines of the same rib, the result is approximated as the straight line in figure 2.7(d). This will eventually result in a cosine error. As shown in figure 2.7(e), the rotation angle is corrected using the slope of the approximate straight line. (4) The pitch value is considered as the distance between two adjacent peak positions. Several hundred pitch values are obtained, and the average of these values is taken as the pitch value at the measurement point.

2.2.2 Analysis of variance (ANOVA)

The measurand in the pitch measurement of the one-dimensional grating is the average pitch. It is not realistic to measure the entire sample surface of a few mm square using a metrological AFM. Usually, several measurement locations are selected, and the measurement is repeated a few times at each measurement location to obtain the average pitch. The main sources of uncertainty in pitch calibration of one-dimensional gratings are the non-uniformity of a specimen and measurement repeatability. Analysis of variance (ANOVA) is used to separate the two factors (Misumi *et al* 2008). Details of ANOVA can be found in general textbooks on statistics (for instance, Winer *et al* 1991). The calculation procedure is as follows. This experiment is a single-factor experiment. As shown in figure 2.8(a), the factors are L and R, and their levels are n_L and n_m. Here, n_L is 10 and n_m is 2. Figure 2.8(b) shows the table of the average pitch p_{ij} ($i = 1, \ldots, n_L$, $j = 1, \ldots, n_m$) at each measurement at each measurement location. The sum of the squared deviations of between-class variation S_A and the sum of the squared deviations of within-class variation S_E are calculated using the following equations,

$$S_A = \sum_{i=1}^{n_L}\sum_{j=1}^{n_m}(\bar{p}_i - \bar{\bar{p}})^2, \tag{2.1}$$

$$S_E = \sum_{i=1}^{n_L}\sum_{j=1}^{n_m}(p_{ij} - \bar{p}_i)^2 \tag{2.2}$$

where $\bar{p}_i (i = 1, \ldots, n_L)$ is the average pitch of all repeated measurements at each measurement location and p is the average pitch of all pitches. The degree of freedom of between-class variation ϕ_A, the degree of freedom of within-class variation ϕ_E, the bias variance of between-class variation V_A, and the unbiased variance of within-class variation V_E are calculated using the following equations,

$$\phi_A = n_L - 1, \tag{2.3}$$

$$\phi_E = n_L(n_m - 1), \tag{2.4}$$

	A	B	C	D
1	■Control Data■			
2				
3				
4	[Factors]			
5	Name of factor (Y)	L		
6	Levels of factor (Y) n_L	10		
7	Name of factor (X)		R	
8	Levels of factor (X) n_m		2	
9	[Parameter or Random Variables]			
10	Type of factor			
11	Two-way split-plot design			
12	[Assign the factors]			
13		1		
14		E2		
15				

(a) Names and levels of factors

	A	B	C	D	E
1	■ANOVA Data■				
2	p_{ij}		$j=1$	$j=2$	
3			R1	R2	
4	$i=1$	L1	24.856	24.968	
5	$i=2$	L2	24.587	24.749	
6	$i=3$	L3	24.582	24.837	
7	$i=4$	L4	24.759	24.830	
8	$i=5$	L5	24.881	24.930	
9	$i=6$	L6	24.874	24.877	
10	$i=7$	L7	24.827	24.793	
11	$i=8$	L8	24.634	24.817	
12	$i=9$	L9	24.944	24.961	
13	$i=10$	L10	24.060	24.530	
14					

(b) Average pitches at each duplicate measurement and measurement location

	A	B	C	D	E	F
1	■Result of ANOVA■					
2						
3						
4		Factor	S	f	V	
5						
6		CF	12265.89766	1	12265.89766	
7						
8		L	S_A 0.65315853	ϕ_A 9	V_A 0.07257317	
9		error2	S_E 0.18317454	ϕ_E 10	V_E 0.01831745	
10						
11		ST	0.83633307	19		
12						

(c) Results of ANOVA

	A	B	C
1	■Expectation of Variance (Completely Randomized Design)■		
2			
3			
4			
5	L $E(V_A)$	$\sigma e_2^2 + 2\sigma L^2$	
6	E2 $E(V_E)$	σe_2^2	
7			

(d) Expectation of variance

Figure 2.8. Procedures of analysis of variance (ANOVA) (Misumi *et al* 2008).

$$V_A = S_A/\varphi_A, \tag{2.5}$$

$$V_E = S_E/\varphi_E. \tag{2.6}$$

The calculation results of equations (2.1)–(2.6) are shown in figure 2.8(c). Next, the expectation of variance is considered. The model of the pitch of each measurement at each measurement location is expressed as follows,

$$p_{ij} = \mu + \alpha_i + \varepsilon_{ij}, \tag{2.7}$$

where μ is the population mean, $\alpha_i(i = 1, \ldots, n_L)$ is the deviation caused by measurement locations, and $\varepsilon_{ij}(i = 1, \ldots, n_L, j = 1, \ldots, n_m)$ is the deviation caused by repetition. The expectation of the sum of the squared deviations of between-class variation $E(S_A)$ is expressed by the following equation,

$$E(S_A) = E\left[\sum_{i=1}^{n_L}\sum_{j=1}^{n_m}(\alpha_i - \bar{\alpha})^2 + \sum_{i=1}^{n_L}\sum_{j=1}^{n_m}(\bar{\varepsilon}_i - \bar{\bar{\varepsilon}})^2\right], \tag{2.8}$$

where α is the average of the deviations caused by measurement locations, $\bar{\varepsilon}_i$ is the average of the deviations caused by repetition at each measurement location, and ε is the average of the deviations caused by repetition. The population standard deviation caused by measurement locations, σ_A, and the population standard deviation caused by duplicate measurements, $\sigma_E(= \sigma_{e2})$, are expressed as the following equations,

$$\sigma_A{}^2 = E\left[\sum_{i=1}^{n_L}(\alpha_i - \bar{\alpha})^2/(n_L - 1)\right], \tag{2.9}$$

$$\sigma_E{}^2 = E\left[\sum_{i=1}^{n_L}(\varepsilon_{ij} - \bar{\bar{\varepsilon}})^2/(n_L n_m - 1)\right]. \tag{2.10}$$

Therefore, $E(S_A)$ is expressed using equations (2.8)–(2.10) as follows,

$$E(S_A) = n_m(n_L - 1)\sigma_A{}^2 + (n_L - 1)\sigma_E{}^2. \tag{2.11}$$

The expectation of the sum of the squared deviations of within-class variation $E(S_E)$ is expressed using equation (2.10) as follows,

$$E(S_E) = E\left[\sum_{i=1}^{n_L}\sum_{j=1}^{n_m}(\varepsilon_{ij} - \bar{\varepsilon}_i)^2\right] = n_L(n_m - 1)\sigma_E{}^2. \tag{2.12}$$

The expectation of variance of between-class variation $E(V_A)$ and the expectation of variance of within-class variation $E(V_E)$ are expressed using equations (2.3), (2.4), (2.5), (2.6), (2.11), and (2.12) as follows,

$$E(V_A) = \sigma_E{}^2 + n_m\sigma_A{}^2, \tag{2.13}$$

$$E(V_E) = \sigma_E{}^2. \tag{2.14}$$

The estimated standard deviation caused by locations $\hat{\sigma}_A$ and the estimated standard deviation caused by duplicate measurements $\hat{\sigma}_E$ are expressed as follows,

$$\hat{\sigma}_A = \sqrt{(V_A - V_E)/n_m}, \tag{2.15}$$

$$\hat{\sigma}_E = \sqrt{V_E}. \tag{2.16}$$

The estimated population mean p is expressed as follows,

$$\begin{aligned}\bar{\bar{p}} &= \sum_{i=1}^{n_L}\sum_{j=1}^{n_m} p_{ij}/n_L n_m = \sum_{i=1}^{n_L}\sum_{j=1}^{n_m}(\mu + \alpha_i + \varepsilon_{ij})/n_L n_m \\ &= \mu + \sum_{i=1}^{n_L}\alpha_i/n_L + \sum_{i=1}^{n_L}\sum_{j=1}^{n_m}\varepsilon_{ij}/n_L n_m.\end{aligned} \tag{2.17}$$

Then, the uncertainty is expressed as follows,

$$u^2(\bar{\bar{p}}) = \hat{\sigma}_A{}^2/n_L + \hat{\sigma}_E{}^2/n_L n_m. \tag{2.18}$$

Therefore, the standard uncertainty caused by non-uniformity of a specimen u_L and the standard uncertainty caused by the repeatability of measurements u_R are expressed as follows,

$$u_L = \hat{\sigma}_A/\sqrt{n_L}, \tag{2.19}$$

$$u_R = \hat{\sigma}_E/\sqrt{n_m \cdot n_L}. \tag{2.20}$$

The results of equations (2.1)–(2.6) are automatically calculated using an ANOVA tool such as Microsoft Excel. The expectation of variance is also needed to calculate uncertainty using ANOVA. However, as shown in figure 2.8(d), the structure of the equation of the expectation of variance changes in the number of factors and levels. Therefore, general ANOVA tools are insufficient to calculate uncertainty. The AIST-ANOVA software developed by NMIJ statisticians was used to calculate the uncertainty as shown in figure 2.8(a)–(d). The AIST-ANOVA software is an add-in software based on the Microsoft Excel ANOVA tool and automatically shows the equation of expectation of variance. The equation for the obtained expectation of the variance is shown in figure 2.8(d). The Japanese version of the AIST-ANOVA software can be downloaded for free.

2.3 Pitch calibration of one-dimensional grating by optical diffractometers

2.3.1 Laser diffractometers

Optical diffractometers are widely used for pitch calibration of one-dimensional gratings in addition to metrological AFMs. Optical diffractometers have the advantage of high throughput. On the other hand, when lasers are used as light sources, the pitch that can be calibrated is limited by the laser wavelength, and a relatively wide periodic structure (mm order) is required. The pitch is obtained from the laser wavelength and the diffraction angle. Therefore, it is necessary that the diffraction angle measured by the rotary encoder built in the rotary table and the laser wavelength have traceability to national standards of length and angle, respectively.

Various laser diffractometers have been developed by NMIs (Korotkov *et al* 1996, Meli *et al* 1999, Kitta *et al* 2004, Buhr *et al* 2007, Pekelsky *et al* 2007). A laser diffractometer in PTB is introduced as an example (Buhr *et al* 2007, figure 2.9).

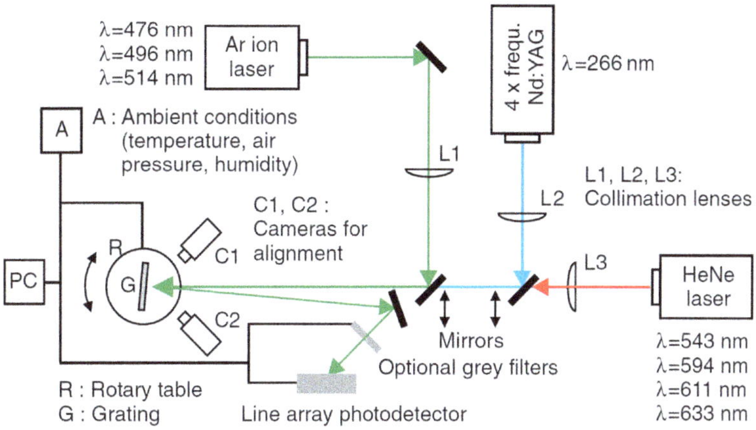

Figure 2.9. Schematic drawing of a laser diffractometer system (Buhr *et al* 2007).

A He–Ne laser, an Ar ion laser and a Nd:YAG laser are mounted as light sources. Eight wavelengths can be selected from 266 nm to 633 nm. The surface of the one-dimensional grating is irradiated with a laser beam from any of the laser light sources through (a mirror and) semi-transparent mirror(s). Diffracted light is detected by a line array photodirector. The diffraction angle is measured by a rotary encoder installed in the rotary table.

Comparisons of pitch measurements between metrological AFM(s) and laser diffractometer(s) have been performed (Meli 2000, Misumi *et al* 2003b, 2007b, 2017). The consistency of measurement results has been confirmed.

2.3.2 Traceable X-ray diffractometer

The minimum pitch that can be calibrated by laser diffractometers is approximately 97 nm (Kitta *et al* 2004, Misumi *et al* 2006). With advances in nanotechnology, there is a need for a method that can calibrate smaller pitches faster. X-ray diffractometers have been developed since the wavelength of X-rays is shorter than wavelengths of lasers (Ito *et al* 2007, Misumi *et al* 2012). Figure 2.10 shows a schematic drawing of the traceable X-ray diffractometer. Figure 2.10(a) is a top view and figure 2.10(b) is a front view. A laser beam of a laser diffractometer irradiates a one-dimensional grating surface perpendicularly. On the other hand, the X-ray beam of the traceable X-ray diffractometer irradiates a one-dimensional grating surface at an incident angle φ as shown in figure 2.10(b). The wavelength of the X-ray source is SI traceable since it is calibrated by Standard Reference Material® 2000 of the National Institute of Standards & Technology (NIST SRM 2000) (National Institute of Standards & Technology, Certificate of Analysis 2012). The angle measured by a rotary encoder built into the rotary table is calibrated, therefore the diffraction angle is also SI traceable. The X-ray diffractometer has a smaller beam diameter (μm order) than the laser diffractometer (mm order). In a one-dimensional grating consisting of a multilayer film structure with a periodic structure region of only a few μm, the pitch

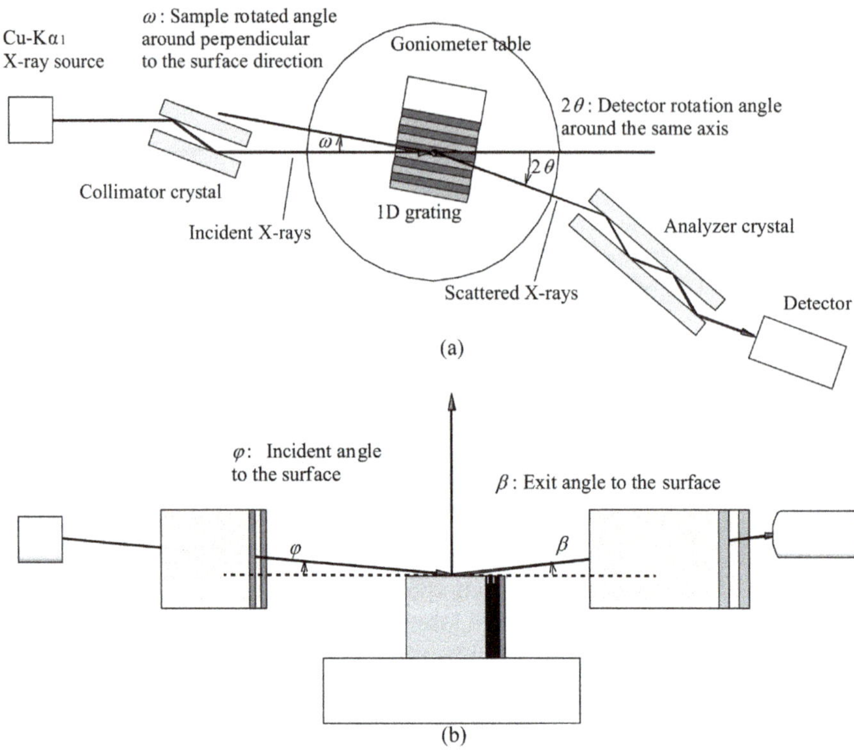

Figure 2.10. Schematic drawing of a traceable X-ray diffractometer: (a) top view; (b) side view (Misumi *et al* 2012).

can be measured by using a traceable X-ray diffractometer. As a result of the pitch comparison measurement between the NMIJ's metrological AFM and the traceable X-ray diffractometer, the consistency of the measurement results has been confirmed (Misumi *et al* 2012, 2017).

2.4 Summary

Firstly, several fabrication methods of one-dimensional gratings were introduced. Next, the pitch calibration method using a metrological AFM was explained. Furthermore, a pitch calibration method using optical diffractometers was also introduced.

References

Buhr E, Michaelis W, Diener A and Mirandé W 2007 Multi-wavelength VIS/UV optical diffractometer for high-accuracy calibration of nano-scale pitch standards *Meas. Sci. Technol.* **18** 667–74

Dai G, Koenders L, Pohlenz F, Dziomba T and Danzebrink H-U 2005 Accurate and traceable calibration of one-dimensional gratings *Meas. Sci. Technol.* **16** 1241–9

ISO 11952 2019 Surface Chemical Analysis—Scanning-Probe Microscopy—Determination of Geometric Quantities Using SPM: Calibration of Measuring Systems

Ito Y, Inaba K, Omote K, Wada Y and Ikeda S 2007 Characterization of submicron-scale periodic grooves by grazing incidence ultra-small-angle X-ray scattering *Jpn. J. Appl. Phys.* **46** L773–5

Itoh H, Fujimoto T and Ichimura S 2006 Tip characterizer for atomic force microscopy *Rev. Sci. Instrum.* **77** 103704

Kitta J, Kubota F and Mine H 2004 Calibration of one-dimensional diffraction grating with an optical diffraction pitch calibration apparatus *Proc. of First Int. Symp. on Standard Materials and Metrology for Nanotechnology (SMAM-1) (Ariake, Tokyo, Japan)* 119–25

Korotkov V I, Pulkin S A, Vitushkin A L and Vitushkin L F 1996 Laser interferometric diffractometry for measurements of diffraction grating spacing *Appl. Opt.* **35** 4782–86

McClelland J J, Anderson W R, Bradley C C, Walkiewicz M and Celotta R J 1993 Laser-focused atomic deposition *Science* **262** 877–80

McClelland J J, Scholten R E, Palm E C and Celotta R J 2003 Accuracy of nanoscale pitch standards fabricated by laser-focused atomic deposition *J. Res. Natl. Inst. Stand. Technol.* **108** 99–113

Meli F 2000 CCL-S1 (NANO4) Final Report https://bipm.org/utils/common/pdf/final_reports/L/S1/CCL-S1.pdf (accessed 21 January 2020)

Meli F and Thalmann R 1998 Long range AFM profiler used for accurate pitch measurements *Meas. Sci. Technol.* **9** 1087–92

Meli F, Thalmann R and Blattner P 1999 High precision pitch calibration of gratings using laser diffractometry *Proc. of the 1st Int. Conf. on Precision Engineering and Nanotechnology* ed P PcKeown *et al(May 1999)* (*Bremen)* 2 252–5

Misumi I *et al* 2003b Submicrometre-pitch intercomparison between optical diffraction, scanning electron microscope and atomic force microscope *Meas. Sci. Technol.* **14** 2065–74

Misumi I *et al* 2007b Round-robin measurements of 100- and 60-nm scales among a deep-ultraviolet laser diffractometer, a scanning electron microscope and various atomic force microscopes *Meas. Sci. Technol.* **18** 803–12

Misumi I *et al* 2012 25 nm pitch comparison between a traceable x-ray diffractometer and a metrological atomic force microscope *Meas. Sci. Technol.* **23** 015002

Misumi I, Dai G, Lu M, Sato O, Sugawara K, Gonda S, Takatsuji T, Danzebrink H-U and Koenders L 2010 Bilateral comparison of 25 nm pitch nanometric lateral scales for metrological scanning probe microscopes *Meas. Sci. Technol.* **21** 035105

Misumi I, Gonda S, Kurosawa T and Takamasu K 2003a Uncertainty in pitch measurements of one-dimensional grating standards using a nanometrological atomic force microscope *Meas. Sci. Technol.* **14** 463–71

Misumi I, Gonda S, Sato O, Sugawara K, Yoshizaki K, Takatsuji T, Azuma Y, Fujimoto T and Kurosawa T 2007a Application of a GaAs/InGaP superlattice in nanometric lateral scales *Meas. Sci. Technol.* **18** 2743–9

Misumi I, Gonda S, Sato O, Sugawara K, Yoshizaki, Kurosawa T and Takatsuji T 2006 Nanometric lateral scale development using an atomic force microscope with directly traceable laser interferometers *Meas. Sci. Technol.* **17** 2041–7

Misumi I, Kitta J, Kizu R and Hirai A 2017 Re-evaluation of calibration and measurement capabilities of pitch calibration systems designed by using the diffraction method *Int. J. Autom. Technol.* **11** 691–98

Misumi I, Lu M, Tanaka H, Sugawara K, Gonda S and Kurosawa T 2008 Nanometric lateral scale development with Si/SiO_2 multilayer thin-film structures and improvement of uncertainty evaluation using analysis of variance *Meas. Sci. Technol.* **19** 045101

Nakayama Y 2005 Highly accurate grating reference fabricated by EB cell projection lithography *Jpn. J. Appl. Phys.* **44** 5586–89

Nakayama Y, Okazaki S and Sugimoto A 1988 Proposal for a new submicron dimension reference for an electron beam metrology system *J. Vac. Sci. Technol.* B **6** 1930–33

National Institute of Standards & Technology, Certificate of Analysis 2012 Standard Reference Material® 2000, Calibration Standard for High-Resolution X-Ray Diffraction https://www-s.nist.gov/srmors/certificates/2000.pdf

Pekelsky J R, Eves B J, Nistico P R and Decker J E 2007 Imaging laser diffractometer for traceable grating pitch calibration *Meas. Sci. Technol.* **18** 375–83

Senoner M, Wirth T, Unger W, Escher M, Weber N, Funnemann D and Krömker B 2005 Testing of lateral resolution in the nanometre range using the BAM-L002 - certified reference material: application to *ToF-SIMS IV* and *NanoESCA* instruments *J. Surf. Anal* **12** 78–82

Winer B J, Brown D R and Michels K M 1991 *Statistical Principles in Experimental Design* 3rd edn (Boston, MA: McGraw-Hill)

Chapter 3

Step height

Step height is another one of the most important nano dimensional standards. The calibration method by metrological AFMs based on ISO 5436-1 is explained. Atomic step is also introduced.

3.1 Definition of step height

Like one-dimensional grating, step height is also one of the most important nano dimensional standards. The calibrated step height standard is used to calibrate the Z-axis of nanometrological instruments. Conventionally, step height standards have been calibrated using stylus instruments. The calculation method of step height is defined in ISO 5436-1 (2000). Figure 3.1 shows the definition of step height in ISO 5436-1 (2000). Figure 3.1 is a cross-sectional view of a line pattern, and the line width of the line pattern is W (Misumi *et al* 2006). The scanning range is set to $3W$ centering on the line pattern. The base straight line is obtained from the left and right $2W/3$ data parts, excluding the edge part $W/3$ of the line pattern. The step height value from the base straight line is calculated from the data of $W/3$ near the center on the line pattern. The data in the vicinity of the edge of the line pattern is not used in the calculation because the measured data in the edge portion cannot be normally acquired. The step height in the AFM measurement is defined in ISO 11952. ISO 11952 is based on ISO 5436-1 (2000) and is extended to surfaces as well as line profiles.

3.2 Step height measurement based on ISO 5436-1 by metrological AFM

Comparison measurements of step height standards were performed by national metrology institutes (NMIs) (Koenders *et al* 2003). The step height standards provided by PTB were used in the comparison (Bienias *et al* 1998, Koenders *et al* 2003).

doi:10.1088/978-0-7503-3191-3ch3

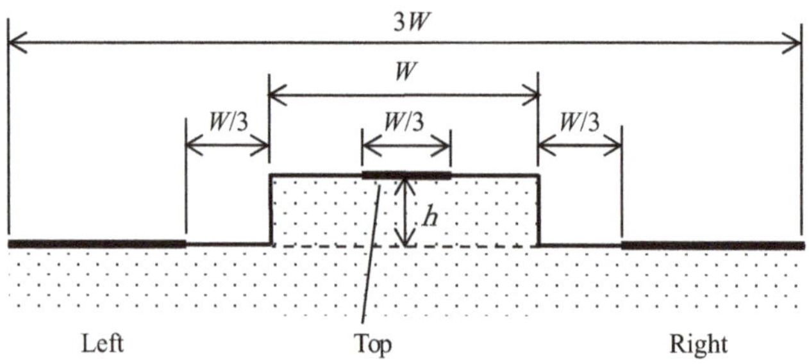

Figure 3.1. Schematic representation of criteria for evaluation of step height standards (ISO 5436-1: 2000). Reprinted from Misumi *et al* 2006. Copyright (2006), with permission from ASPE, EUSPEN and JSPE.

Figure 3.2(a) shows an optical micrograph of a step height standard. Line patterns of SiO_2 are formed on a silicon substrate. Five types of step height standards with different heights between 6 nm and 800 nm were prepared. A chromium thin film is formed on a standard surface to protect the surface of the standard and to make the optical constants of the surface the same. There were three types of line patterns for one type of step height standard, while the width of the line pattern used for the comparison measurement was 30 μm. In addition to metrological AFMs, stylus instruments and optical instruments were also used for comparison measurements. Figure 3.2(b) shows an example of a metrological AFM image of the step height standard captured by NMIJ. The definition of step height in this comparison measurement is determined as an analogy of ISO 5436-1 since the scanning ranges of metrological AFMs are not wide in many cases. The definition shown in figure 3.3 was used in the comparison measurement. Figure 3.4 shows an example of comparison measurement results. It seems that coating with Cr was effective. In this comparison measurement, the consistency of the measurement results was confirmed for almost all NMIs during different measurement methods (Koenders *et al* 2003).

3.3 Range of reliability of measurement data

ISO 5436-1 (2000) recommends setting the data range of the reference straight line on both sides of the line pattern as shown in figure 3.1. However, due to the shape of the sample and the limitations of AFMs, it may not be possible to secure a sufficient scanning range for the reference straight line portion as in the international comparison in section 3.2. Also, in some cases, such as an atomic step, the reference straight line may be obtained only on one side. Why is it recommended to set the data range of the reference line on both sides of the line pattern? The reliability of the parameters of the Gaussian-related basic straight line can be calculated using the least-squares method, and the difference in reliability between the step height standard based on ISO 5436-1 (2000) and an ordinal atomic step structure can be compared (Misumi *et al* 2006).

The calculation procedure of the reliable range at any position of the related feature $\mathbf{S}_m$ is described. The following calculations assume that cross-section features have

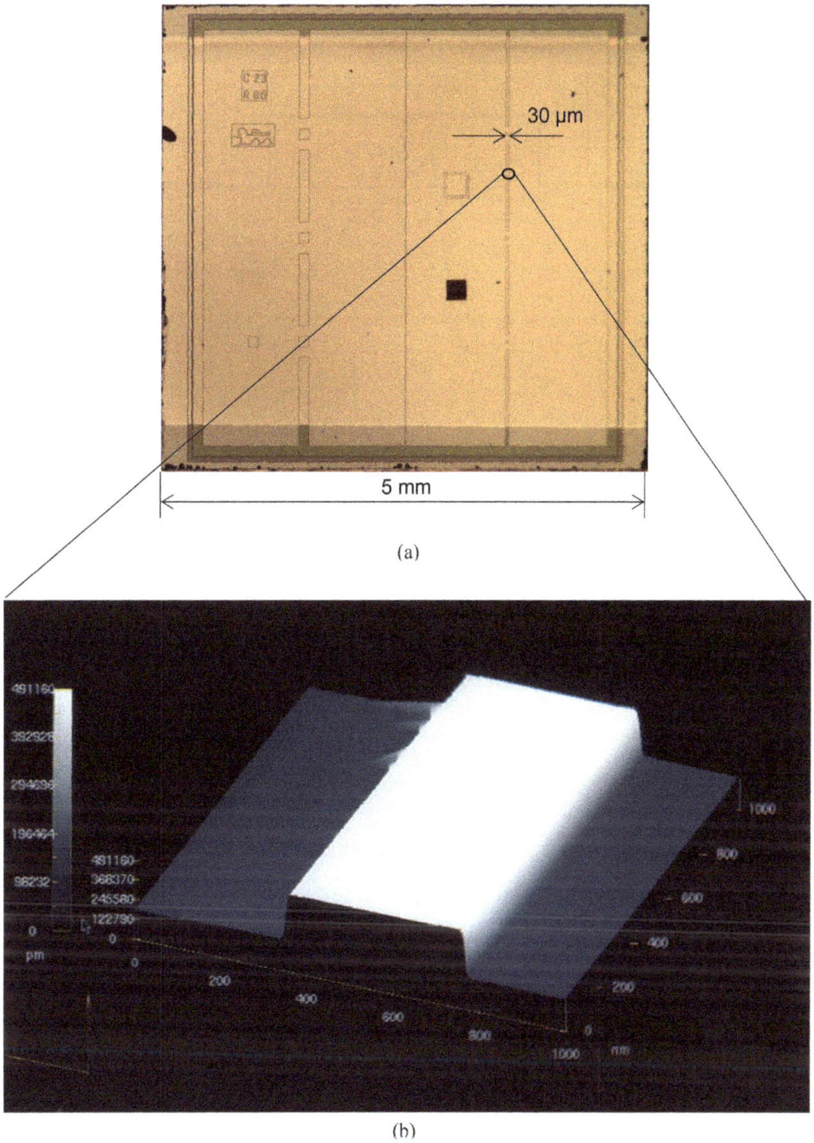

Figure 3.2. (a) Optical microscope image of a step height standard. (b) Metrological AFM image of the step height standard captured by NMIJ.

random errors, are uncorrelated, and have a normal distribution. The observation equation (3.1), the ordinary equation (3.2), and the least-squares solution (3.3) have the following relationship:

$$\mathbf{d} = \mathbf{A}\mathbf{p}, \tag{3.1}$$

$$\tilde{\mathbf{A}}\mathbf{S}^{-1}\mathbf{A}\mathbf{p} = \tilde{\mathbf{A}}\mathbf{S}^{-1}\mathbf{d}, \tag{3.2}$$

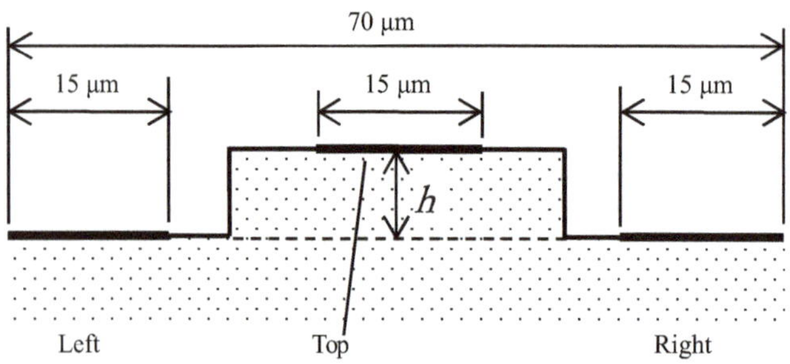

Figure 3.3. Definition of step height in the comparison measurement.

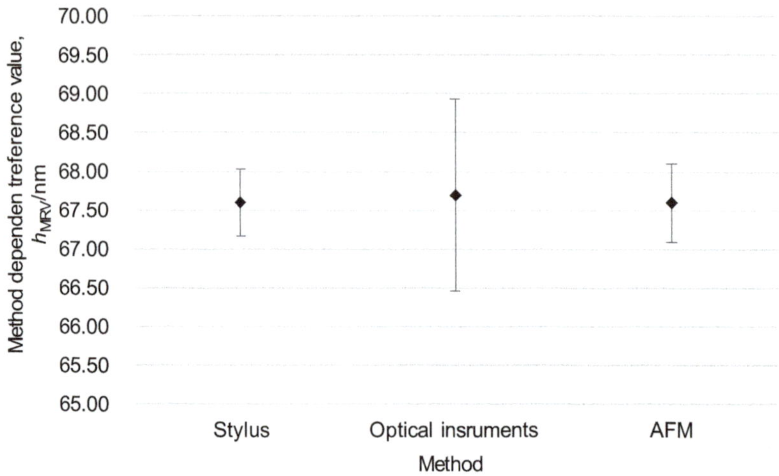

Figure 3.4. An example of the method dependent reference value of the step height standard.

$$\mathbf{p} = (\tilde{\mathbf{A}}\mathbf{S}^{-1}\mathbf{A})^{-1}\tilde{\mathbf{A}}\mathbf{S}^{-1}\mathbf{d}, \tag{3.3}$$

where **d** is the measured data vector, **A** is the Jacobian matrix, **p** is the parameter vector, **S** is the error matrix of the measured data, and $\tilde{\mathbf{A}}$ is the transposed matrix of **A**. The error matrix of the parameter $\mathbf{S}_p$ and the calculated value $\mathbf{S}_m$ are obtained as shown in equations (3.4) and (3.5). The diagonal elements of $\mathbf{S}_p$ indicate the error variance of the parameters, and $\mathbf{S}_m$ indicates the reliable range variance of the Gaussian-related features at any point on the related features. Equation (3.4) represents the transfer from the error in the measured data **S** to the error in the parameter $\mathbf{S}_p$. Equation (3.5) represents the transfer from the error of the parameter $\mathbf{S}_p$ to the error of the calculated value $\mathbf{S}_m$.

$$\mathbf{S}_p = (\tilde{\mathbf{A}}\mathbf{S}^{-1}\mathbf{A})^{-1}, \tag{3.4}$$

$$\mathbf{S}_m = \mathbf{A}\mathbf{S}_p\tilde{\mathbf{A}}. \tag{3.5}$$

If the error in the measured data **S** is known, the error range (confidence range) $\mathbf{S}_m$ at any point of the relevant feature can be calculated using the Jacobian matrix. The Jacobian matrix depends only on the location of the measured points. Therefore, the error of the parameter $\mathbf{S}_p$ can be evaluated from the error of the position of the measured point and the measurement data **S**. The reliability of the calculation result can be evaluated by using the least-squares method. From the above equation, we can calculate the reliable range at any position of the related feature $\mathbf{S}_m$. The procedure for calculating the reliability of the parameters of the basic straight line is as follows. The parameters p_1 and p_2 are determined from the y-intercept and the slope. Then from the observation equation, equation (3.1) becomes $d = p_1 + p_2 x$. Based on the measured points t_1 $(x_1, y_1),\ldots, t_n$ (x_n, y_n), the linear Jacobian matrix **A** is expressed as

$$\mathbf{A} = \begin{pmatrix} 1 & x_1 \\ \vdots & \vdots \\ 1 & x_n \end{pmatrix}. \tag{3.6}$$

Next, the error matrix for the parameter $\mathbf{S}_p$ is obtained. The measured points are placed symmetrically with respect to the origin of the region between $x = -0.5D$ and $x = 0.5D$, the measured points are uncorrelated, and the variance of the measurement data is assumed to be $\sigma_0{}^2$. Then, the error matrix for the parameter $\mathbf{S}_p$ is expressed as

$$\mathbf{S}_p = \begin{pmatrix} \sigma_{p1}{}^2 & 0 \\ 0 & \sigma_{p2}{}^2 \end{pmatrix} = (\tilde{\mathbf{A}}\mathbf{S}^{-1}\mathbf{A})^{-1} = \sigma_0{}^2 \begin{pmatrix} 1/n & 0 \\ 0 & 1/\sum x_i{}^2 \end{pmatrix}, \tag{3.7}$$

where n is the number of measured points and x_i is the position of the measurement data.

The associated base line is on the X-axis. Therefore, the Jacobian matrix of line **A** can be easily expressed as

$$\mathbf{A} = (1 \quad x). \tag{3.8}$$

The reliable range at any position of the associated base line is expressed as

$$\mathbf{S}_m = \sigma_m{}^2 = \sigma_0{}^2\left(1/n + x_i{}^2/\sum x_i{}^2\right). \tag{3.9}$$

As shown in figure 3.5(a), the position of the measured points (number: $2n$) on the base straight line is assumed. Half of the measured points n are evenly spaced in the region between $x = D$ and $x = lD$ and the remaining points n are in the region between $x = -D$ and $x = -lD$. In this case, the normalized reliable range σ_m/σ_0 is expressed as

$$\sigma_m/\sigma_0 = \sqrt{1/2n + (1/(n + (l-1)(n+1) + (n+1)(2n+1)(l-1)^2/6n)) \cdot x^2/D^2}. \tag{3.10}$$

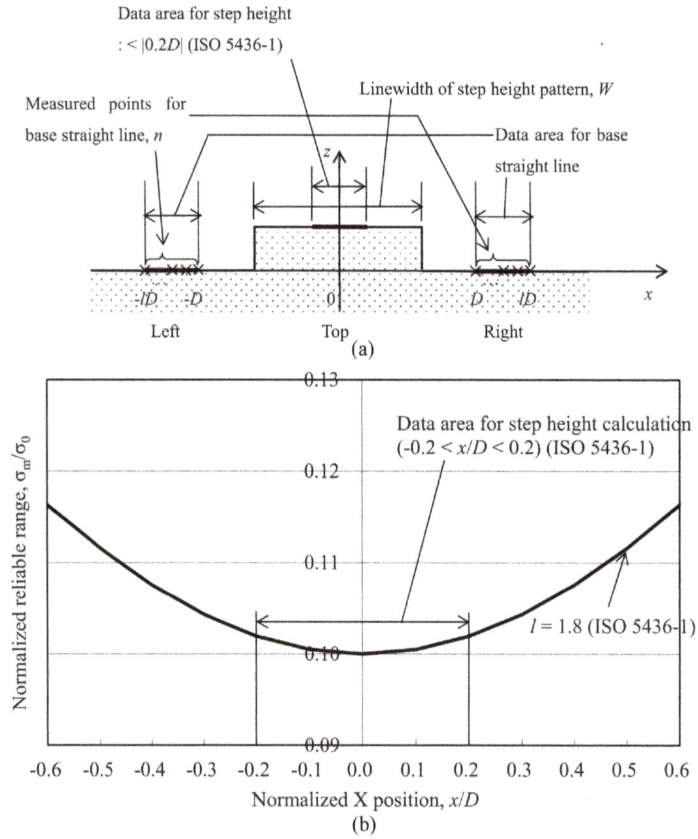

Figure 3.5. (a) Measured points for base straight line, *n* in the case of a step height standard based on ISO 5436-1 (2000). (Republished with permission of Elsevier Science & Technology Journals, from Misumi I *et al* 2006 *Precis. Eng.* **30** 13–22; permission conveyed through Copyright Clearance Center, Inc.) (b) Normalized reliable range σ_m/σ_0 vs. normalized *X* position in various areas of measured data for associated base straight line *l*. Reprinted from Misumi *et al* 2006. Copyright (2006), with permission from ASPE, EUSPEN and JSPE.

Therefore, equation (3.10) can be used to evaluate the reliable range of the base line. Figure 3.5(b) shows the relationship between the normalized reliable range σ_m/σ_0 and the normalized horizontal position $x = D$ (number of measured points, $2n = 100$). The solid line indicates the normalized reliable range within the measured data of the base line of $l = 1.8$ (ISO 5436-1: 2000). Step heights are calculated in the range $|x/D| < 0.2$. The maximum normalized confidence range σ_m/σ_0 for $|x/D| < 0.2$ is approximately 0.103. This range is added to the measurement uncertainty evaluation as one of uncertainty components.

For an ordinal atomic step sample, the positions of the measured points (number: *n*) on the base line are set as shown in figure 3.6(a). The atomic step sample is assumed to have the same terrace widths, *W*. The base straight line is calculated from the measured points between $-0.5D$ and $0.5D$. The step height is calculated using data

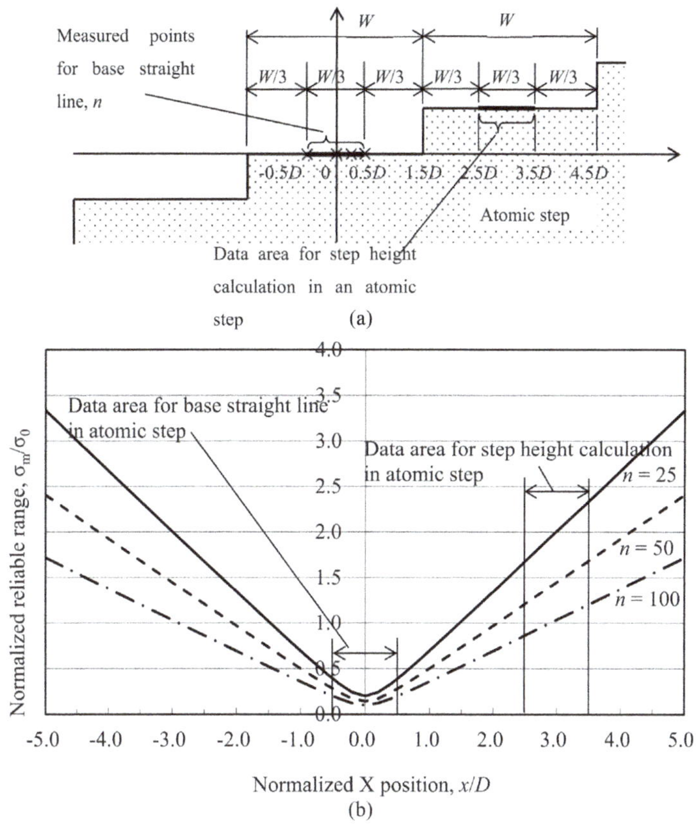

Figure 3.6. (a) Measured points for base straight line, n in the case of an ordinal atomic step sample. (b) Normalized reliable range σ_m/σ_0 vs. normalized X position in various numbers of measured data for associated base straight lines. Compared with figure 3.5(b), the normalized reliable range is significantly reduced. Reprinted from Misumi *et al* 2006. Copyright (2006), with permission from ASPE, EUSPEN and JSPE.

between 2.5D and 3.5D. The data area for the step height calculation and the data area for the base straight line calculation are equal to $W/3$. From equation (3.9) and figure 3.6(a), the normalized confidence range σ_m/σ_0 is expressed as the following equation,

$$\sigma_m/\sigma_0 = \sqrt{1/2n + ((12 + (n - 1))/(n(n + 1))) \cdot x^2/D^2}. \tag{3.11}$$

Figure 3.6(b) shows the calculation results of the normalized confidence range σ_m/σ_0 for various measured points n on the base straight line. The normalized reliable range between 2.5D and 3.5D was about 0.8–2.3. These values were much larger than those for the normal step height sample shown in figure 3.5(b). This is because the data area for calculating the step height of an ordinal atomic step is not in the center region of the data area for the base straight line calculation. A reliable range of parameters for the base straight line can be one of the major sources of uncertainty.

In order to measure step height standards accurately, the definition of ISO should be adhered to as much as possible, the line pattern whose height is to be measured should be placed in the center of the scanning range, and the base straight line should be obtained from the measurement results on both sides of the line pattern.

3.4 Atomic step

The heights of most step height standards fabricated by semiconductor lithography technology are several nanometers or more. Atomic step samples with sub-nanometer height have been developed. Figure 3.7(a) shows a schematic diagram of a conventional silicon atomic step. Homma *et al* developed Si atomic steps in 1990 (Homma *et al* 1990). Furthermore, Suzuki *et al* promoted the standardization of the Si atomic step measurement method (Suzuki *et al* 1996). At NIST, Si atomic steps were calibrated using the C-AFM (Tsai *et al* 1998, Orji *et al* 2004). Other NMIs have also tried calibrating atomic steps with metrological AFMs. However, it is challenging to calibrate atomic step samples with a sufficiently small uncertainty. This is due to the fact that if the step height becomes smaller, uncertainty caused by metrological AFMs, such as the periodic error of the interferometric measurement, becomes relatively large.

There is, however, a movement to calibrate a commercially available AFM by using atomic step samples. In 2019, guidelines on how to evaluate uncertainty in measuring atomic steps based on bulk lattice constants were issued (Koenders *et al* 2019). In order to compensate for the drawbacks of the Si atomic step pointed out in section 3.3, the development of a silicon atomic step having a so-called amphitheater structure is also in progress (Yacoot *et al* 2007, Garnaes *et al* 2020). Figure 3.7(b) shows a schematic diagram of the Si atomic step having an amphitheater structure.

The AFM image of the amphitheater Si atomic step is shown in figure 3.8(a) (Garnaes *et al* 2020). In particular, an AFM image taken by the commercially available AFM is affected by the non-linear error $\delta z_e(x)$ such as drift. It is shown as

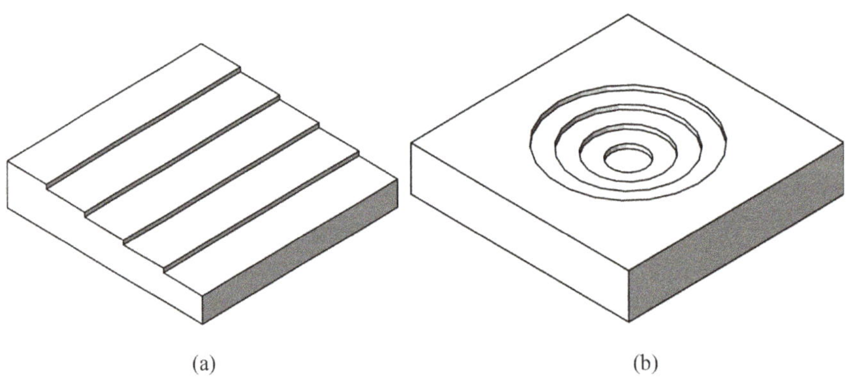

Figure 3.7. Schematic drawing of Si atomic step samples: (a) an ordinal silicon atomic step, (b) a so-called amphitheater structure of a silicon atomic step.

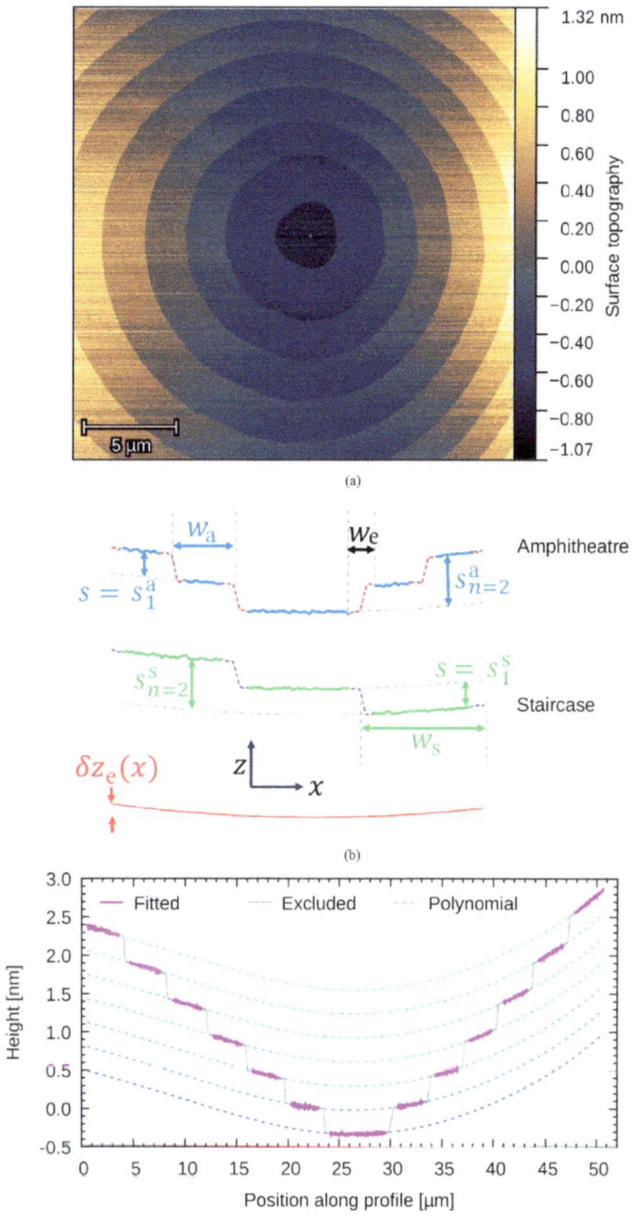

Figure 3.8. (a) AFM topographic image of a monoatomic step silicon amphitheater structure. (b) Diagram of the shape of the stairs and the amphitheater. (c) Step height fitting using fourth order polynomial background removal. The dashed lines represent the background of common polynomials that shifted to individual terrace levels (Garnaes *et al* 2020). Copyright 2020 BIPM and IOP Publishing Ltd. CC BY 4.0.

$\delta z_e(x)$ in figure 3.8(b) (Garnaes *et al* 2020). Using a two-straight-parallel-line fit near the same edge will result in a different step height than the spacing between the grid planes. The effects of non-linear errors in the surface profile $\delta z_e(x)$ (see figure 3.8(b)) are eliminated in the fitting process. The appropriate section of profile width near the edge is excluded from the analysis using medium-width terraces w_a and w_s for the amphitheater and stairs, respectively (see figure 3.8(b)). Due to the small sub-nm (0.314 nm) height of the monoatomic steps, AFM data processing procedures such as line leveling, background removal, and statistical analysis are important for accurate results. This is shown in figure 3.8(c). Here, a high-order polynomial fit is used to subtract the non-linear background variation. In addition, advanced data analysis routines are required to further improve fitting results, and curve-optimized segmentation is required to avoid adverse effects on edge and step height measurements. During data fitting, the height of the steps should be kept constant across multiple steps.

Figure 3.9 shows a flowchart of the height evaluation procedure for the Si atomic step (Garnaes *et al* 2020). The height of a step is evaluated by fitting the surface with a piecewise function that combines the step with a smooth polynomial background. This procedure has two basic parts: dividing the image into individual terraces and least-squares fitting with a surface model. Segmentation can be performed in a variety of ways, from fully automated steps to manual step tracking. It is assumed that the individual terraces are properly marked and fitted.

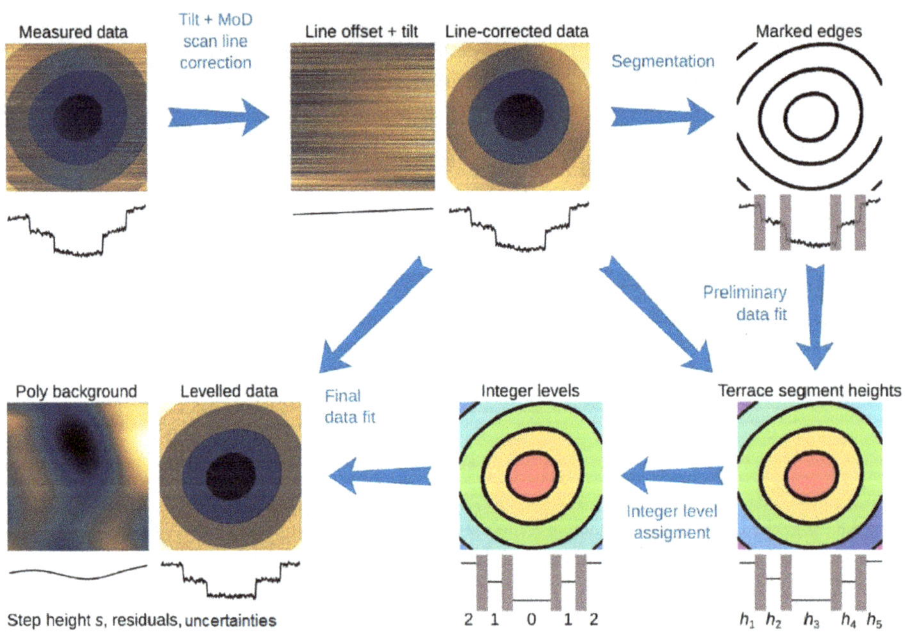

Figure 3.9. Flowchart of the step height evaluation procedure. Each group of images and profiles represents the inputs and outputs of individual steps and is shown as blue arrows. The black outline of the image and the brown bar of the profile represent the marked and masked edges (Garnaes *et al* 2020). Copyright 2020 BIPM and IOP Publishing Ltd. CC BY 4.0.

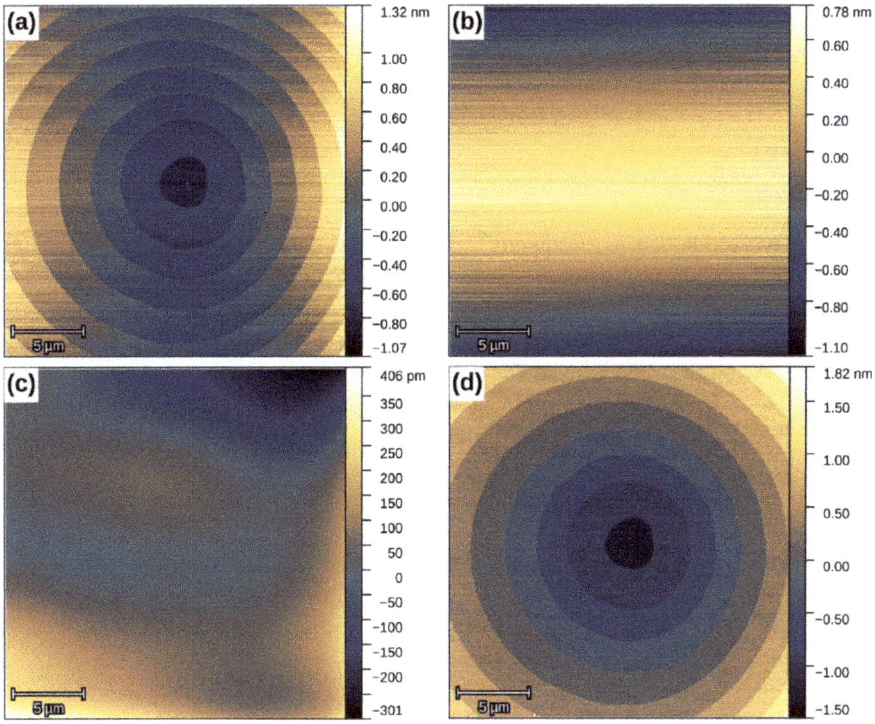

Figure 3.10. Comparison of (a) the original measured data, (b) the deducted scanline mismatch (b), (c) the polynomial background, and (d) the data after removing both backgrounds. Image (a) is the sum of (b), (c), and (d). The image that fits the step height evaluation is the sum of (c) and (d) (Garnaes *et al* 2020). Copyright 2020 BIPM and IOP Publishing Ltd. CC BY 4.0.

With this method, it is necessary to find the background of the global polynomial and the height of each terrace at the same time. It is pre-fitted without making any assumptions about the height of individual terrace segments. Next, the integer level is assigned to the terrace segment. A final fit is made with a known terrace level that results in a step height h.

Since the 2D analysis provides a background for the 2D polynomial, it can also be compared to the original data as shown in figure 3.10 (Garnaes *et al* 2020) (this is the same measurement as in figure 3.8). In addition, the third component in the data (a value that is not part of the polynomial background of the approximate model of the step) can be isolated. These data are associated with random defects in the scan. As can be seen in figure 3.10(b), this random part is relatively prominent and is associated with a typical defect in AFM microscopes, which is a mismatch between individual lines. Fast scan axis data (usually individual profiles that appear horizontally) are usually very consistent, but individually in relation to mechanical and thermal drifts, tip shape changes, or feedback loop performance changes, there is often a slight shift between the profiles.

3.5 Summary

Step height calibration based on the definition of ISO 5436-1 (2000) using a metrological AFM was explained. Atomic step height measurements were also introduced.

References

Bienias M, Gao S, Hasche K, Seemann R and Thiele K 1998 A metrological scanning force microscope used for coating thickness and other topographical measurements *Appl. Phys.* A **66** S837–42

Garnaes J, Necas D, Nielsen L, Madsen M H, Torras-Rosell A, Zeng G, Klapetek P and Yacoot A 2020 Algorithms for using silicon steps for scanning probe microscope evaluation *Metrologia* **57** 064002

Homma Y, McClelland J and Hibino H 1990 DC-resistive-heating-induced step bunching on vicinal Si (111) *Jpn. J. Appl. Phys.* **29** L2254–6

ISO 5436-1 2000 Geometrical Product Specifications (GPS)—Surface Texture: Profile Method; Measurement Standards—Part 1: Material Measures

Koenders L *et al* 2003 Comparison on nanometrology: nano 2—step height *Metrologia* **40** 04001

Koenders L, Busch I, Garnaes J, Yacoot A, Dixson R and Bosse H 2019 Recommendations of CCL/WG-N on: Realization of SI metre using height of monoatomic steps of crystalline silicon surfaces *Guidance document CCL-GD-MeP*-3

Misumi I, Gonda S, Kurosawa T, Azuma Y, Fujimoto T, Kojima I, Sakurai T, Ohmi T and Takamasu K 2006 Reliability of parameters of associated base straight line in step height samples: uncertainty evaluation in step height measurements using nanometrological AFM *Precis. Eng.* **30** 13–22

Orji N G, Dixson R G, Fu J and Vorburger T V 2004 Traceable pico-meter level step height metrology *Wear* **257** 1264–9

Suzuki M *et al* 1996 Standardized procedure for calibrating height scales in atomic force microscopy on the order of 1 nm *J. Vac. Sci. Technol.* A **14** 1228–32

Tsai V W, Vorburger T, Dixson R, Fu J, Köning R, Silver R and Williams E D 1998 The study of silicon stepped surfaces as atomic force microscope calibration standards with a calibrated AFM at NIST *AIP Conf. Proc.* **449** 839–42

Yacoot A, Koenders L and Wolff H 2007 An atomic force microscope for the study of the effects of tip–sample interactions on dimensional metrology *Meas. Sci. Technol.* **18** 350–9

Ichiko Misumi

Chapter 4

Two-dimensional grating

Pitch calibration of a two-dimensional grating has a difficulty that is different from that of a one-dimensional grating. The pitch calibration method of two-dimensional grating by metrological AFM is explained.

4.1 Two-dimensional grating

It is possible to prepare two one-dimensional gratings and install them so that the line pattern directions are in the *X*- and *Y*-axis directions, respectively, in order to calibrate the *XY* axes of the nanometrological instruments. However, there is a possibility that two standard samples cannot be installed if the measurement chamber of the nanometrological instrument is small. Just one standard sample can be used to calibrate both the *XY* axes of the nanometrological instrument if a two-dimensional grating is installed (Chernoff et al 2008).

International comparison measurements of two-dimensional gratings were performed between national metrology institutes (NMIs) (Garnaes and Dirscherl 2008). Two types of instruments, metrological AFMs and optical diffractometers, are used in thc comparison, which is the same as the comparison of a one-dimensional grating (Garnaes and Dirschcrl 2008, Buhr *ct al* 2007, Dai *et al* 2007, Pan *et al* 2011). As shown in figure 4.1, there are three measurands in the comparison: the average pitch of the column, p_x, the average pitch of the row, p_y, and the counterclockwise angle θ between the row direction and the column direction. The mathematical model of the two-dimensional grating is as follows from figure 4.1.

$$\cos\theta = \left(h_x{}^2 + h_y{}^2 - h_{xy}{}^2\right)/2h_xh_y, \tag{4.1}$$

$$\theta = \theta_1 + \theta_2, \tag{4.2}$$

doi:10.1088/978-0-7503-3191-3ch4

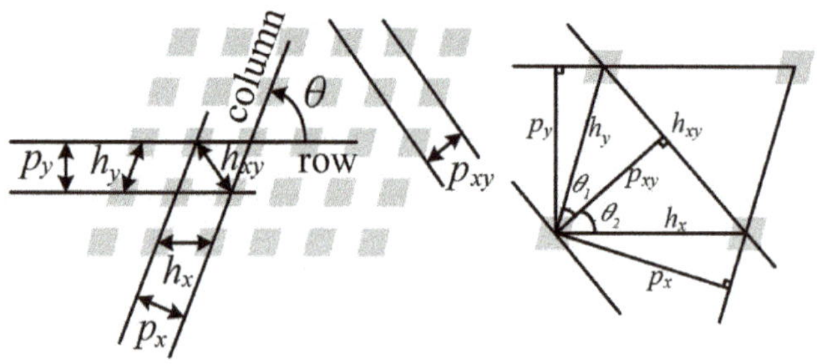

Figure 4.1. The measurands of the 2D grating calibrations: the mean pitches p_x, p_y, and the orthogonality α in the anticlockwise direction (Pan *et al* 2011).

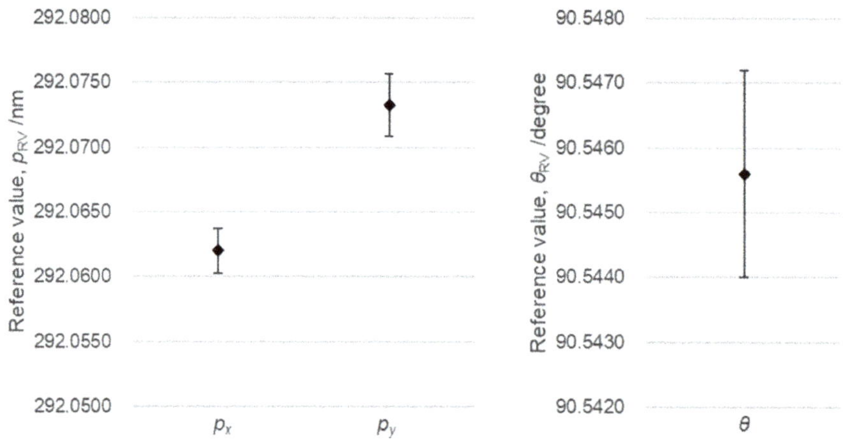

Figure 4.2. An example of the reference values in the comparison measurements of two-dimensional gratings.

$$p_x = h_x \sin\theta = h_{xy} \cos\theta_1, \tag{4.3}$$

$$p_y = h_y \sin\theta = h_{xy} \cos\theta_2, \tag{4.4}$$

$$p_{xy} = h_x \cos\theta_2 = h_y \cos\theta_1. \tag{4.5}$$

From equations (4.1)–(4.5), the following equation is obtained,

$$\cos\theta = \left(p_x^2 + p_y^2 - p_x^2 p_y^2 / p_{xy}^2\right) / 2 p_x p_y. \tag{4.6}$$

The angle θ can be calculated if p_x, p_y, and p_{xy} can be obtained from measurement data. Figure 4.2 shows an example of international comparison results. The method dependent reference values have not been calculated in this comparison. The results obtained by the optical diffraction method and the metrological AFM method were calculated collectively.

4.2 Pitch calibration of two-dimensional grating by a metrological AFM

In general, AFM data has high reliability in the fast scan axis, but not in the slow scan axis. This is because drift during AFM measurements cannot be avoided. The reliability of the slow scan axis is not high even with the metrological AFMs. It is necessary to calculate the pitches of p_x, p_y, and p_{xy} from the three different metrological AFM images by changing the scanning directions of p_x, p_y, and p_{xy} three times in order to calibrate the two-dimensional grating more precisely. Aligning the scanning direction with the p_x, p_y, and p_{xy} directions is a laborious and difficult task.

Dai *et al* of PTB proposed a method for adjusting the fast scan axis in the p_x and p_y directions. Figure 4.3 shows the measurement strategy for calibrating a two-dimensional grating. The two-dimensional grating is scanned over three different measurement areas. Area A is a small square area for displaying the 2D grating in detail. Areas B and C are two narrow rectangular areas to determine the average pitch and local pitch uniformity along the *X*- and *Y*-axes, respectively. Figure 4.4(a) shows how to determine the fast scan direction and figure 4.4(b) shows an example of three profiles measured along different scan directions. θ in figure 4.4(b) is the angle between the scan direction and the *X*-axis of the instrument coordinate system. By finely adjusting the scanning direction so that the line profile showing the periodic structure can be obtained, the pitch can be calculated more quickly.

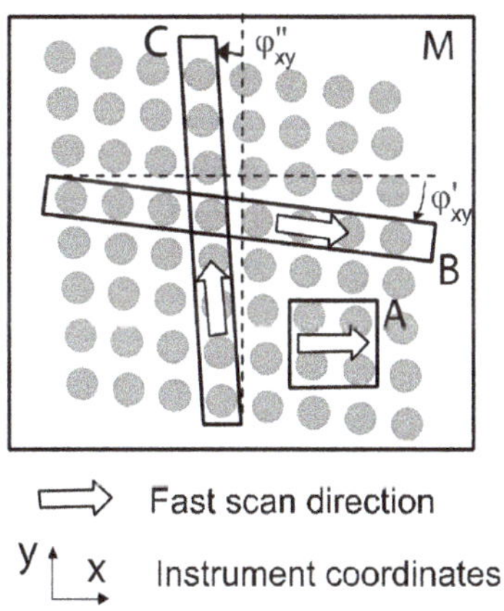

Figure 4.3. Schematic diagram illustrating a new measurement strategy for calibrating two-dimensional gratings. The grating is scanned over three different measurement areas. Area A is a small square area for viewing grating grids in detail. Areas B and C are two narrow rectangular areas for determining the mean pitch and local pitch uniformity along the *X*- and *Y*-axes, respectively (Dai *et al* 2007). Copyright IOP Publishing Ltd. All rights reserved.

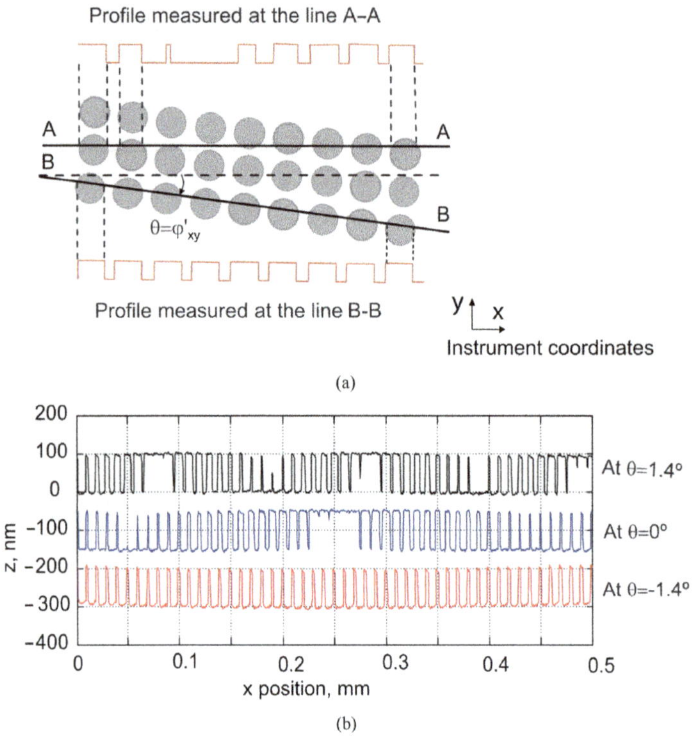

Figure 4.4. (a) Schematic diagram illustrating the method for determining the fast scan direction, and (b) three profiles measured along different scan directions (Dai *et al* 2007).

4.3 Summary

Firstly, the measurand of two-dimensional gratings was explained. The pitch calibration of two-dimensional grating using a metrological AFM was also introduced.

References

Buhr E, Michaelis W, Diener A and Mirandé W 2007 Multi-wavelength VIS/UV optical diffractometer for high-accuracy calibration of nano-scale pitch standards *Meas. Sci. Technol.* **18** 667–74

Chernoff D A, Buhr E, Burkhead D L and Diener A 2008 Picometer-scale accuracy in pitch metrology by optical diffraction and atomic force microscopy *Proc. of SPIE* **6922** 69223J

Dai G, Pohlenz F, Dziomba T, Xu M, Diener A, Koenders L and Danzebrink H-U 2007 Accurate and traceable calibration of two-dimensional gratings *Meas. Sci. Technol.* **18** 415–21

Garnaes J and Dirscherl K 2008 NANO5—2D grating—final report *Metrologia* **45** 04003

Pan S-P, Liu T-S, Tasi M-C and Liou H-C 2011 Method for determining the angle in two dimension nanoscale: pitch grating *Jpn. J. Appl. Phys.* **50** 06GJ05

Chapter 5

Line width

Calibration of line width (or critical dimension (CD)) is greatly affected by the atomic force microscope (AFM) tip shape. Special metrological AFMs (CD-AFMs and tilted-type AFMs) for line width calibration are explained. The calibration method by a combination of a metrological AFM and transmission electron microscope (TEM) is also shown.

5.1 Critical dimension atomic force microscope (CD-AFM)

It is possible to define measurands that are not affected by the probe-tip shape in one-dimensional grating, step height, and two-dimensional grating measurements. On the other hand, the measurand of the line width (or CD) has to be considered as being influenced by the probe-tip shape. CD-AFM is one of the AFMs often used for CD measurement. A flared CD probe-tip is used in a CD-AFM, and the scanning of the probe-tip is controlled in two axial directions. The CD-AFM is suitable for measuring sidewalls of line patterns.

Figure 5.1(a) shows an ordinal AFM probe-tip path, and figure 5.1(b) shows a CD-AFM probe-tip path, respectively (Orji and Dixson 2007). In a conventional AFM, the apparent width of the structure includes the extension due to the probe-tip as a function of height. In a CD-AFM, most of the expansion of the apparent width due to the flared probe-tip is composed of a constant term (zero-order component). If the width of the flared probe-tip is known, the line width can be obtained by subtracting the width of the probe-tip from the apparent profile.

If the bottom surface of the flared CD probe-tip is perfectly flat, the expansion of the apparent width due to the probe-tip is only a constant term. However, the actual CD probe-tip is rounded. Figure 5.2 shows examples of actual flared CD probe-tips (Orji and Dixson 2007). The CD probe-tips have shapes as shown in figures 5.2(a)–(c),

doi:10.1088/978-0-7503-3191-3ch5

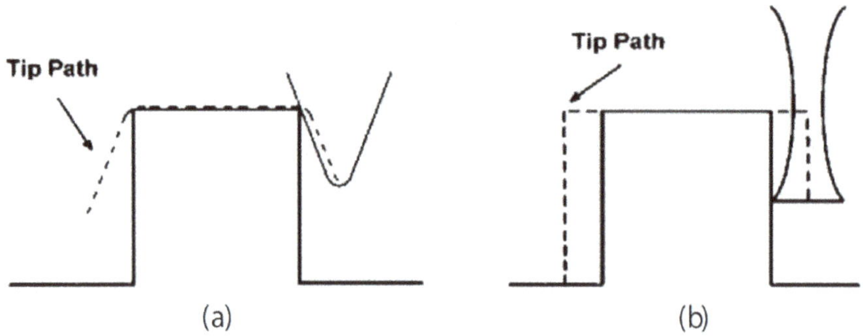

Figure 5.1. Diagram of the difference between (a) a conventional AFM and (b) a CD-AFM. In a conventional AFM, the apparent width of the structure includes the extension due to the probe-tip as a function of height. In a CD-AFM, most of the extension of the apparent line width due to the flared CD probe-tip is a constant term (the zeroth-order component) (Orji and Dixson 2007).

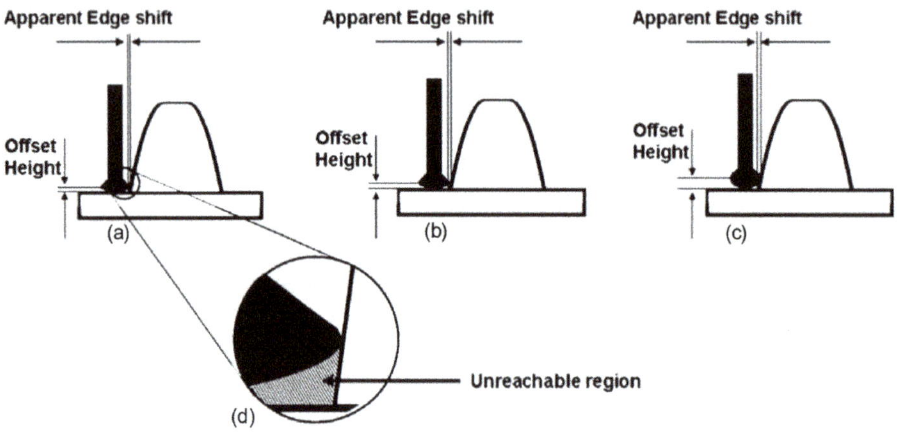

Figure 5.2. (a)–(c) Flared CD probe-tips with different offset heights in contact with the feature sidewalls. (d) An unreachable region due to offset height (Orji and Dixson 2007).

and have different offset heights. As shown in figure 5.2(d), the offset height creates an unreachable region. This unreachable region must be taken into consideration when obtaining the sidewall angle of the line pattern.

Figure 5.3(a) shows a CD-AFM image of isolated line features of photoresist (Orji and Dixson 2007). Figure 5.3(b) shows a typical profile of the image, showing the features being undercut by an average of 7.6 nm. The total edge overhang of the probe-tip used in this image is 11 nm, so the re-entrant sidewalls can be perfectly captured. The top width of the line pattern was 46.4 nm ± 1.07 nm ($k = 1$) down 20% from the top, but the center width was 37.8 nm ± 1 nm ($k = 1$) after correcting the 111.5 nm flared CD probe-tip width. These uncertainty statements include components from higher order tip effects and account for tip-to-tip variability.

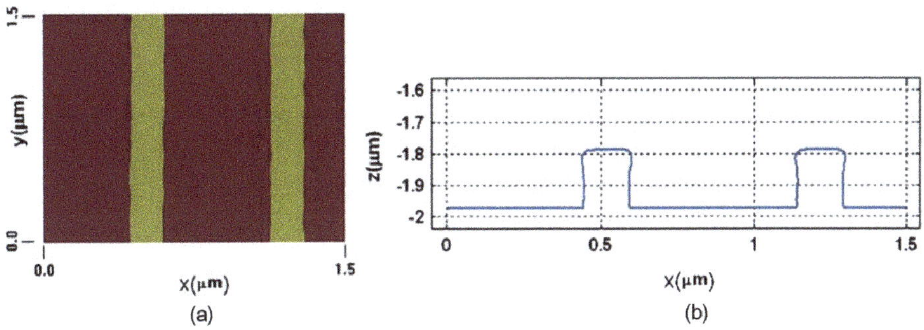

Figure 5.3. (a) AFM image of isolated photoresist line features. (b) Profile from the image (a). The re-entrant nature of the profile is shown (Orji and Dixson 2007). Copyright IOP Publishing Ltd. All rights reserved.

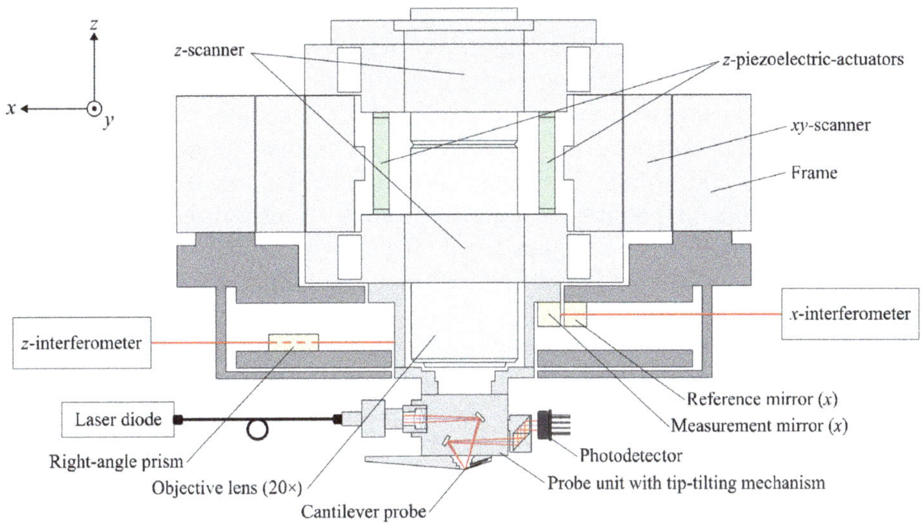

Figure 5.4. Cross-sectional view of the main unit of the tilting-mAFM, which is composed of a 3D scanner, a probe unit with a tip-tilting mechanism, including an optical lever system, laser interferometer modules, and an optical microscope (Kizu *et al* 2018). Copyright IOP Publishing Ltd. All rights reserved.

5.2 Metrological AFM with a tip-tilting mechanism

A CD-AFM, which uses a flare CD probe-tip, has the advantage of being able to accurately measure sidewall shapes of features. On the other hand, the diameter of the flared CD probe-tip is larger than that of an ordinal probe-tip, therefore, it is difficult to measure finer shapes. It is necessary to measure the sidewalls of features more accurately by using an ordinal probe-tip to measure CD.

NMIJ has developed a metrological AFM with a tip-tilting mechanism (tilting-mAFM) (Kizu *et al* 2018). Figure 5.4 shows a cross-sectional view of the main unit of the tilting-mAFM. This equipment consists of a 3D scanner, a probe unit, a laser interferometer module for each axis, and an optical microscope. The probe unit includes a tilting probe mechanism and a conventional optical lever system as a force detector with a cantilever probe. In figure 5.4, the y-axis laser interferometer is located behind the body. The tilting-mAFM is a probe scanning type AFM. In addition to conventional chip-size standard samples, large size samples such as wafers and photomasks can be measured using the coarse movement stage.

Figure 5.5(a) shows a probe unit equipped with a tip-tilted mechanism. The probe unit is turned upside down and attached to the 3D scanner. The probe-tip of the tilting-mAFM is tilted along the rotation around the vertical axis, and the torsional vibration of the cantilever is used to perform probe control. The tilt angle (θ_t) is manually adjusted within a maximum range of ±16°. The rotation guide is a cylindrical sliding surface whose central axis intersects the top of the probe-tip. The cantilever probe is attached to the probe holder and vibrates by the piezoelectric device. The removable probe holder is attached to the tilt unit using a magnet. The entire optical lever system, including the fiber collimator and four segmented photodetectors (PDs), is integrated into a tilt. The θ_t limit is determined by geometrical parameters such as cantilever length, probe-tip length, cantilever probe-tip geometry, and cantilever probe forward tilt angle. When a commercially available, ordinal cantilever probe is tilted laterally, the shoulder of the tip first hits the surface of the sample, even at small θ_t. To further tilt the probe without interference from the surface, both sides of a commercial cantilever probe-tip are trimmed, and a fine tip is realized. Figure 5.5(b) shows a customized cantilever probe made by Olympus for the tilting-mAFM. This probe is based on the OMCL-AC200TS probe (resonance frequency: 150 kHz (typ), spring constant: 9 N · m^{-1} (typ), tip width: 0.4 mm).

It is necessary to perform two sets of measurements on the left and right sidewalls to measure CD since the tilting-mAFM has only one probe. The CD sample is rotated 180° and is measured around the sidewalls on both sides using the tilted mAFM (Kizu *et al* 2019). The two sets of AFM measurement data are stitched together to reconstruct the shape of the line pattern and the CD is determined, as shown in figure 5.6.

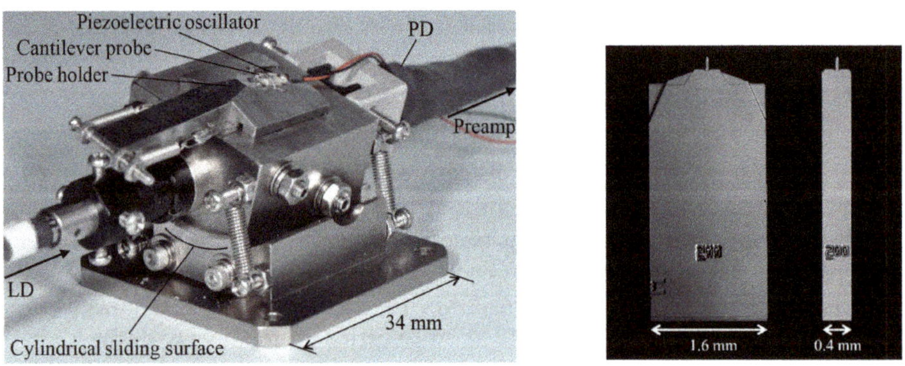

Figure 5.5. (a) The probe unit with tip-tilting mechanism. LD, laser diode; PD, photodetector. (b) Commercial cantilever probe (left) and customized cantilever probe (right) (Kizu *et al* 2018). Copyright IOP Publishing Ltd. All rights reserved.

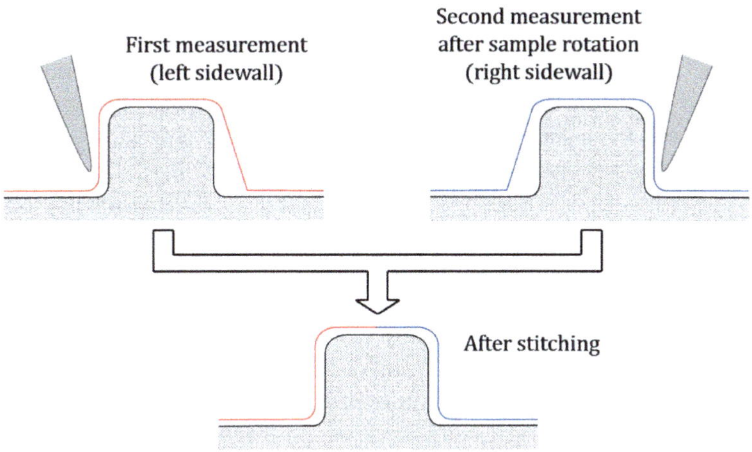

Figure 5.6. Two AFM datasets of the measurements of left and right sidewalls stitched together (Kizu *et al* 2019).

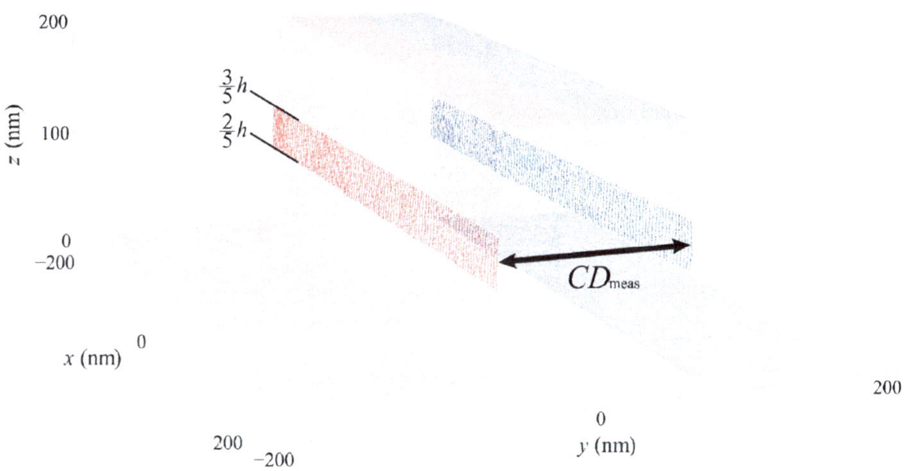

Figure 5.7. Point clouds at heights between $2h/5$ and $3h/5$ selected to perform the CD calculation. These data are highlighted in a deep color (Kizu *et al* 2019).

The CD is calculated when the two sets of the tilting-mAFM measurement data are combined. CD is defined here as the width of the line pattern in the middle of the height. First, the distance between the top and bottom is extracted using the point cloud plane fitting. Measurements in these areas are affected by the convolution of the probe geometry, therefore, the data near the upper corners are excluded. The measurement points are selected at a height between $2h/5$ and $3h/5$ (as shown in figure 5.7) and plane fit is performed. Finally, the central CD value is calculated based on the average distance between two sidewall faces of height $h/2$.

5.3 CD standard sample

Convolution of the probe-tip shape is inevitable in CD measurement, in both a CD-AFM and tilting-mAFM. A CD standard sample for evaluating the probe-tip shape is required. In order to accurately measure the probe width and remove the convolution of the probe shape, NIST developed single crystal critical dimension reference materials (SCCDRM) (Dixson *et al* 2005). The method for realizing the CD reference material is as follows. (1) Single crystal silicon is selectively etched to pattern well-defined and highly uniform features. (2) A high-resolution transmission electron microscope (HRTEM) is used to directly access the Si lattice spacing as a source of SI traceable width information. (3) The CD-AFM is used as a comparator to transfer the width information from the HRTEM sample. An example of a CD-AFM image on the SCCDRM feature is shown in figure 5.8 (Kramar *et al* 2011). The sidewall angle of these structures is very close to 90°.

PTB has also developed a CD standard sample IVPS-100-PTB (Dai *et al* 2015). The IVPS-100-PTB sample was developed based on the traditional commercial sample IVPS-100. Its layout is shown in figure 5.9(a). The sample size is 6 mm × 6 mm. It consists of four independent groups of 5 × 5 functional patterns. This allows the samples to be cut into smaller pieces, for example, by applying some of them to destructive TEM measurements. The layout of the characteristic pattern is shown in figure 5.9(b). It consists of a group of five line features used as reference line features.

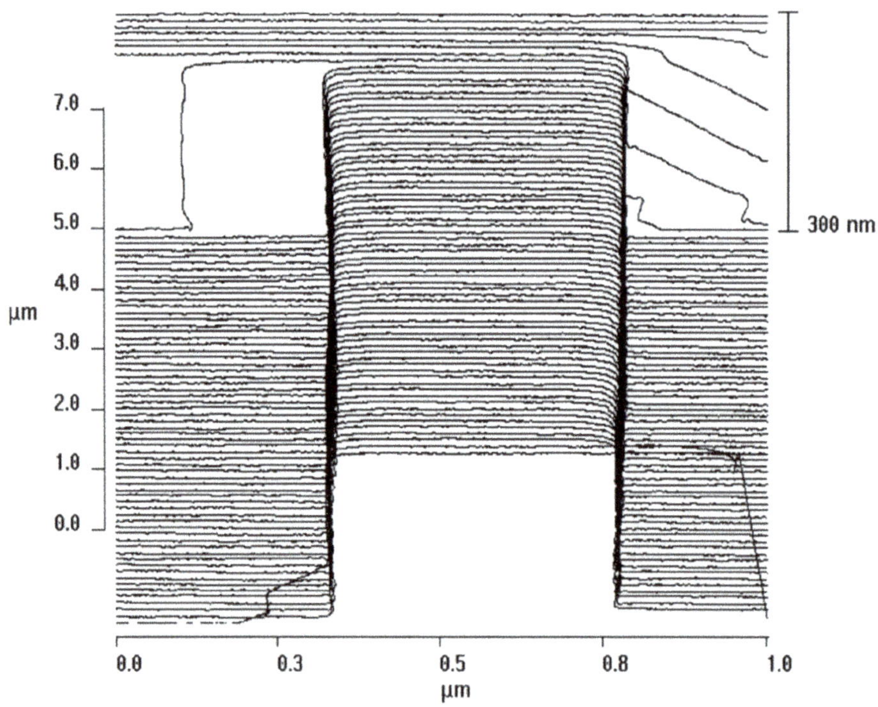

Figure 5.8. An example of a CD-AFM image on the SCCDRM feature (Kramar *et al* 2011). Copyright IOP Publishing Ltd. All rights reserved.

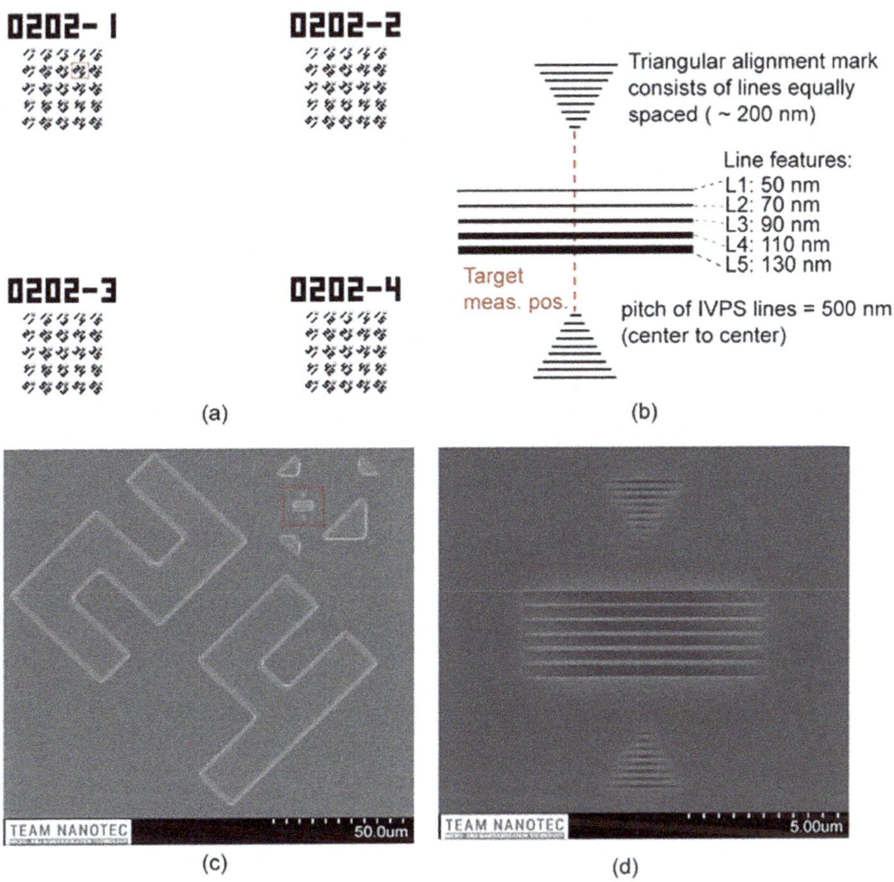

Figure 5.9. (a) A sample-designed layout consisting of four groups of feature patterns. Each pattern has 5 × 5 feature groups. (b) Designed layout of a feature group with two alignment marks and a group of five line features. (C) SEM image of the processed feature group and its label number. (d) SEM image of alignment marks and line features in the marked area of (c) (Dai *et al* 2015). Copyright IOP Publishing Ltd. All rights reserved.

These line features, like the IVPS-100, have vertical parallel sidewalls. The five line features have different nominal CD values of 50 nm, 70 nm, 90 nm, 110 nm, and 130 nm, respectively. The feature pattern contains a pair of triangular alignment marks. The position defined by the two vertices of the mark is the target measurement position. To reduce manufacturing time, the triangular marks are realized as a group of nine equally spaced (~200 nm) grooves. The length of the shortest groove segment is less than 0.4 μm, so it can be properly applied to pinpoint the measurement position. IVPS-100-PTB defines the structure of a masking material manufactured using electron beam lithography and lattice selective wet etching with alkaline solution. Finally, the etch mask is removed and the sample is cleaned before packing in a cleanroom-compatible container. Figure 5.9(c) shows an SEM image showing an overview of the created feature pattern. The labeling number near the reference function helps one to navigate the features of the visual microscope. Figure 5.9(d) shows another SEM image showing details of reference features and alignment marks.

Figure 5.10(a) is a scanning transmission electron microscope (STEM) image of the feature pattern measured at 7100× magnification (Dai *et al* 2015). The figure clearly shows a cross-section of the two groups of alignment marks (each with nine grooves). The nine groove depths are not equal because there is a correlation between the length and depth of the marker alignment groove. This shows that all the grooves of the alignment mark are cross-sectioned in the sample preparation and that the deviation between the actual measurement position and the desired measurement position is less than ±0.2 μm. Five line features L1 … L5 are displayed in the center of the two alignment marks. The purpose of the TEM measurements is to characterize the reference feature CD, corner radius, sidewall angle (SWA), SiO_x layer thickness, and surface quality. Images with different magnifications are acquired for different measurements in order to achieve the best possible measurement performance. For example, the entire line feature is captured at a magnification of 640 kx to 1.3 Mx to

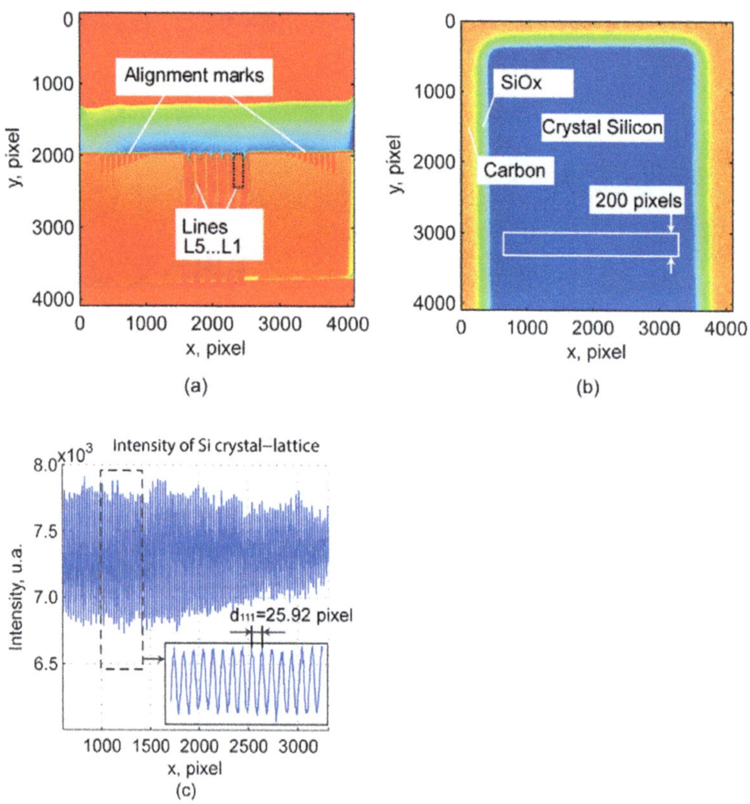

Figure 5.10. (a) An image of alignment marks and five line features measured by STEM (magnification: 7.1 k). (b) Line L1 with nominal feature width of 50 nm measured by STEM (magnification 1.3 M). (c) Average intensity profile of the STEM image in the marked area of (b) with the inset showing the periodic crystal lattice planes of silicon (Dai *et al* 2015). Copyright IOP Publishing Ltd. All rights reserved.

measure CD, and only the corners or sidewalls of the feature are captured at a higher magnification (3.6 Mx) to measure the corner radius. Figure 5.10(b) shows a typical STEM image of line feature L1 (nominal CD 50 nm) taken at 1.3 Mx magnification. The figure presents a very good contrast of the three materials: crystalline silicon, SiO_x, and carbon. The crystal lattice constant of silicon can be applied as an internal scale to calibrate the magnification of (S)TEM measurement, providing high accuracy and traceability. Such a calibration is shown in figure 5.10(c), and the average intensity profile is shown in the marked area of figure 5.10(b). The periodic structure showing the {1 1 1} crystal plane is clearly visible in the inset. Using the Fourier transform, the period d_{111} is calculated as 25.92 pixels. Therefore, with a known value for d_{111} traced to 313.560 11 pm combining X-ray and optical interferometry, the magnification of the TEM image can be determined to be 0.0121 nm/pixel (Massa *et al* 2009).

5.4 Summary

Special metrological AFMs (CD-AFM and tilted-type AFM) for line width calibration were explained. A calibration method by a combination of a metrological AFM and TEM was also introduced.

References

Dai G, Zhu F, Heidelmann M, Fritz G, Bayer T, Kalt S and Fluegge J 2015 Development and characterisation of a new line width reference material *Meas. Sci. Technol.* **26** 115006

Dixson R G, Allen R A, Guthrie W F and Cresswell M W 2005 Traceable calibration of critical-dimension atomic force microscope linewidth measurements with nanometer uncertainty *J. Vac. Sci. Technol.* B **23** 3028–32

Kizu R, Misumi I, Hirai A, Kinoshita K and Gonda S 2018 Development of a metrological atomic force microscope with a tip-tilting mechanism for 3D nanometrology *Meas. Sci. Technol.* **29** 075005

Kizu R, Misumi I, Hirai A, Kinoshita K and Gonda S 2019 Linewidth calibration using a metrological atomic force microscope with a tip-tilting mechanism *Meas. Sci. Technol.* **30** 015004

Kramar J A, Dixson R and Orji N G 2011 Scanning probe microscope dimensional metrology at NIST *Meas. Sci. Technol.* **22** 024001

Massa F, Mana G, Kuetgens U and Ferroglio L 2009 Measurement of the lattice parameter of a silicon crystal *New J. Phys.* **11** 053013

Orji N G and Dixson R G 2007 Higher order tip effects in traceable CD-AFM-based linewidth measurements *Meas. Sci. Technol.* **18** 448–55

IOP Publishing

Nanoscale Standards by Metrological AFM and Other Instruments

Ichiko Misumi

Chapter 6

Nano particle size

Sizes of standard nano particles have been calibrated by several methods, such as the dynamic light scattering (DLS) method, differential mobility analyzer (DMA), and transmission electron microscope (TEM). Recently, the size measurement of nano particles fixed on a substrate has been performed not only by optical microscopes or SEM but also by metrological AFM. Standard nano particles are made from several materials, for instance polystyrene latex (PSL), gold, silver, and silica. Standard PSL nano particles are selected in this chapter, since they have high sphericity.

6.1 Definition and calculation method of nano particle size

For the measurement of nano particle size by AFM, the particle sizes in the height direction and in the lateral direction were defined as measurands that are not affected by the probe-tip shape (Meli 2005, Meli and Küng 2011, Meli *et al* 2012). Figure 6.1(a) shows the definition of nano particle size in the height direction, and figure 6.1(b) shows the definition of nano particle size in the lateral direction. The nano particle size in the height direction is the distance between the substrate and the apex of the nano particle. This definition applies to most types of nano particles. The lateral particle size shown in figure 6.1(b) is a definition that is limited to the types of particles that have a high sphericity and that can make a close-packed structure on the substrate, such as standard PSL particles.

Figure 6.2 shows the calculation procedure of PSL nano particle sizes in the lateral direction (Misumi *et al* 2018). The calculation procedure is an extension of 'the center of gravity method' used to calculate the pitch of a one-dimensional grating. (1) An AFM image is obtained. The scanning position is adjusted so that the area where the nano particles form a close-packed structure is located at the center of the AFM image. Sufficiently large base areas to the left and the right of the particle area are obtained. The left and right base areas serve as a base plane when

doi:10.1088/978-0-7503-3191-3ch6

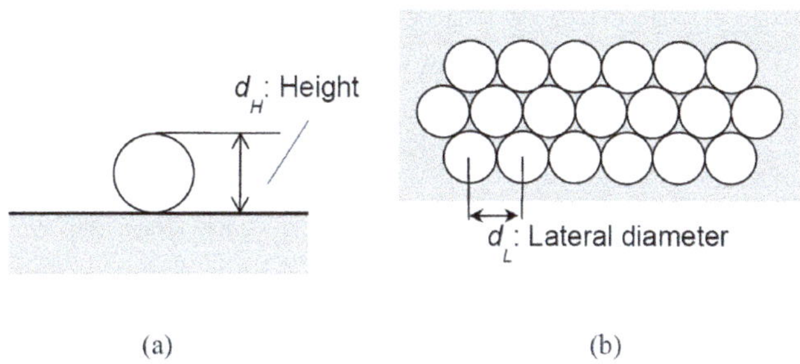

Figure 6.1. Definitions of nano particle size measurements. (a) Particle size in the height direction. (b) Particle size in the lateral direction. Reprinted from Misumi *et al* 2018. Copyright (2018), with permission from ASPE, EUSPEN and JSPE.

nano particle sizes in the height direction are calculated in the same way as a step height measurement. (2) Slope correction is performed using the data of the left and right base plane areas. (3) A threshold plane is set parallel to the base plane. When it is assumed that the base plane, z, as $z = 0$, the height of the threshold plane, z_t is set between the half of a nominal diameter, $d/2$ and a nominal diameter, d, as $d/2 < z_t < d$. The area below $d/2$ of each particle is the area which the AFM probe-tip cannot reach, and the AFM control is the most unstable near $d/2$. A threshold plane is set to remove unnecessary parts including unstable data for particle size calculation. (4) The parts including the apex of each particle and the part surrounded by the threshold plane are cut out. (5) The center of gravity of the cut-out object is calculated. A vertical line is drawn from the center of gravity to the threshold plane. The intersection of the perpendicular and the threshold plane is defined as the peak position. (6) On the threshold plane, the least-squares lines of the peak positions are calculated on the threshold plane as shown in figure 6.2. (7) The particle size in the lateral direction is the distance between neighboring peak positions along the least-squares line.

6.2 Nano particle size measurement using a metrological AFM

Table 6.1 shows the specifications of standard nano particles measured by a metrological AFM at NMIJ. There are three types of particles: PSL 30 nm, PSL 100 nm, and PSL 300 nm. These are the same samples used in the international comparison of nano particle size measurements (APMP.L-S5) (Lin *et al* 2019). The particle sizes were measured by DMA for PSL 30 nm and PSL 100 nm, and by TEM for PSL 300 nm, and the nominal values and their expanded uncertainties are shown.

Figure 6.3 shows an example of a metrological AFM image of PSL 100 nm particles. The substrate is mica. The standard PSL nano particles have a high sphericity and form a close-packed structure. By arranging the particles in the center of the image and the base planes on both sides of the image, the particle diameters in the lateral direction and in the height direction can be calculated. The particle sizes in the lateral direction are calculated only in the region having no air gaps.

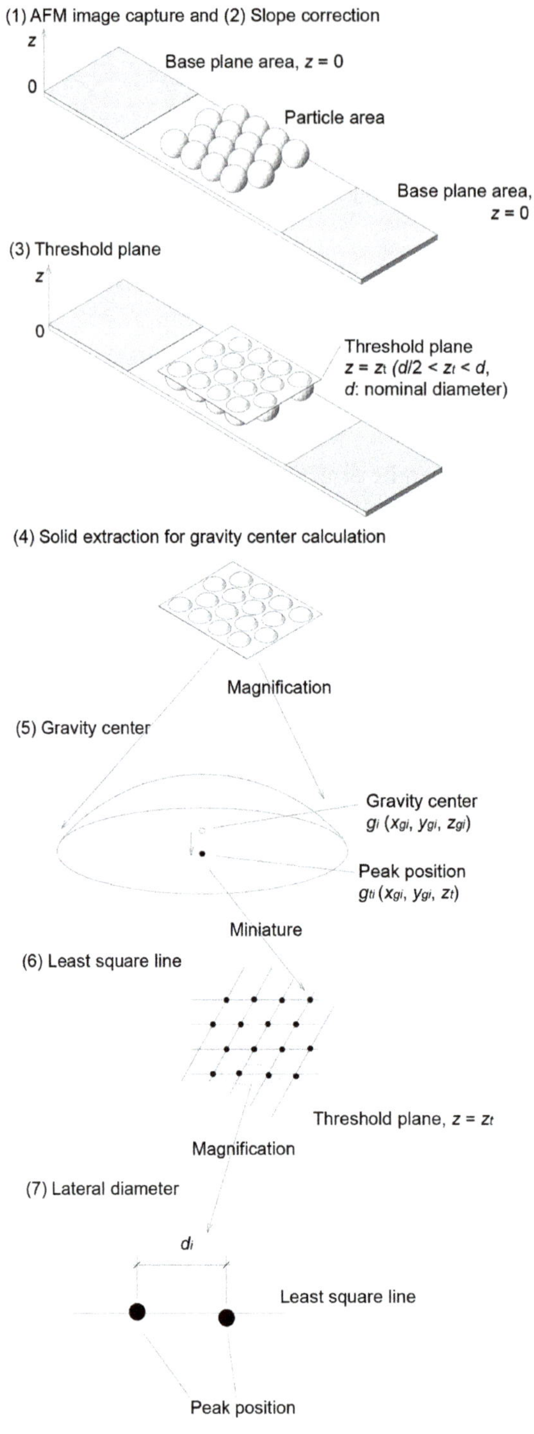

Figure 6.2. The extended gravity center method (GCM). Reprinted from Misumi *et al* 2018. Copyright (2018), with permission from ASPE, EUSPEN and JSPE.

Table 6.1. Specifications of standard PSL nano particles measured by the metrological AFM at NMIJ.

Sample name	Product number	Manufacturer	Nominal diameter, d (nm)	Expanded uncertainty, U (k = 2) (nm)	Particle size measurement method
PSL 30 nm	SC-0030-A	JSR Life Sciences Corporation	29	1	DMA
PSL 100 nm	SC-010-S	JSR Life Sciences Corporation	100	3	DMA
PSL 300 nm	SC-032-S	JSR Life Sciences Corporation	309	9	TEM

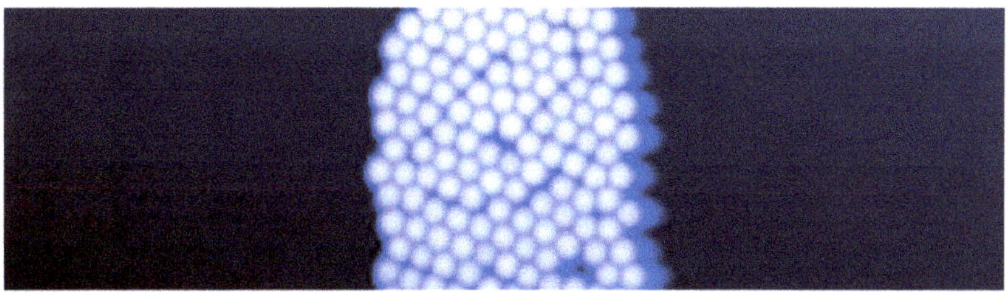

Figure 6.3. AFM image of 100 nm polymer reference particles on mica. In defect-free areas, the average distance between spheres is measured. The apex height of the spheres can be measured relative to the flat mica substrate (black).

Table 6.2 shows the measurement results of the nano particle sizes in the height and lateral directions of PSL 30 nm, 100 nm, and 300 nm and their expanded uncertainties (Misumi *et al* 2018). The particle size is not corrected by particle deformation or air gaps. Also, the uncertainties derived from particle deformation and air gaps are not evaluated.

6.3 Calculation of particle deformation

When the measurement results are compared between AFMs, there are no problems with the measured particle sizes and their expanded uncertainties as explained in section 6.2. However, the size measurement of standard nano particles is performed by various methods (Meli *et al* 2012, Lin *et al* 2019), and therefore it is necessary to correct the deformation of particles when compared with other methods. The elastic deformation model (Garnaes 2011) and plastic deformation model are used to calculate the deformation of particles. The Maugis–Pollock model is applied as one of the methods of assuming plastic deformation in this study (Maugis and Pollock 1984, Rimai *et al* 2000, Wang *et al* 2011, Misumi *et al* 2018). Only the amount of deformation when the particles adhere to the substrate is considered. The amount of deformation of particles due to the probe-tip is considered to be much smaller than the amount of deformation of a particle due to the adhesion between a particle and a substrate, therefore, it is not considered here.

Table 6.2. Measurement results using the metrological AFM at NMIJ. Reprinted from Misumi *et al* 2018. Copyright (2018), with permission from ASPE, EUSPEN and JSPE.

Sample name	Direction	Measured mean diameter, d (nm)	Expanded uncertainty, U_{95} (nm)
PSL 30 nm	Height	23.3	1.3
	Lateral diameter	25.1	1.6
PSL 100 nm	Height	95.4	0.7
	Lateral diameter	97.0	1.9
PSL 300 nm	Height	302.0	0.5
	Lateral diameter	309.9	1.7

Particle size is not corrected by particle deformation or air gap.
Uncertainties derived from particle deformation and air gap are not incorporated.

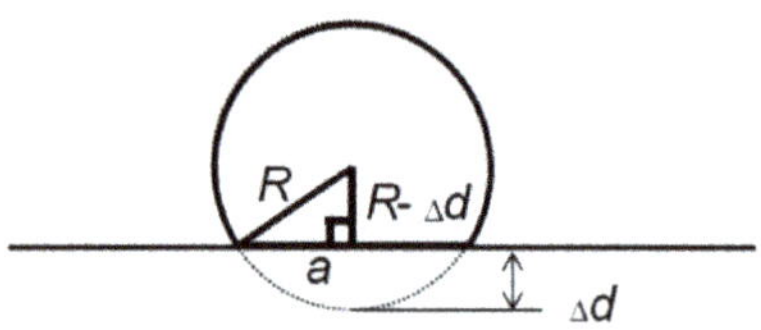

Figure 6.4. Particle deformation model (the Maugis–Pollock model or the MP model); a is the radius of the contact area, R is the radius of the particle, and Δd is the particle deformation. Reprinted from Misumi *et al* 2018. Copyright (2018), with permission from ASPE, EUSPEN and JSPE.

Figure 6.4 shows a schematic drawing of the Maugis–Pollock model. R is the nominal value of the particle radius, a is the radius when the contact area is assumed to be a circle, and Δd is the amount of deformation of the particle. The contact area a is represented by the following formula,

$$a = \sqrt{2w_a R/3Y}, \tag{6.1}$$

where w_a is the work caused by the adhesion between a particle and a substrate, and Y is the yield point of a particle. The work w_a is expressed by the following formula,

$$w_a = \gamma_p + \gamma_s - \gamma_{ps} = \gamma_p + \gamma_s - \left(\gamma_p + \gamma_s - 2\sqrt{\gamma_p \cdot \gamma_s}\right) = 2\sqrt{\gamma_p \cdot \gamma_s}, \tag{6.2}$$

where γ_p and γ_s are the surface energies of the particle and the substrate, respectively. γ_{ps} is the interfacial energy between the particle and the substrate. γ_{ps} is expressed by the following equation,

$$\gamma_{ps} = \gamma_p + \gamma_s - 2\sqrt{\gamma_p \cdot \gamma_s}. \tag{6.3}$$

From equations (6.1)–(6.3), Δd is expressed by the following equation,

$$\Delta d = R - \sqrt{R^2 - a^2} = R - \sqrt{R^2 - 4\sqrt{\gamma_p \cdot \gamma_s} R/3Y}. \tag{6.4}$$

Table 6.3 shows the material properties of PSL and mica, respectively. The amount of particle deformation can be calculated from equation (6.4), the material properties shown in table 6.2, and the nominal particle radius. The reliability of the physical properties is unknown. Therefore, it is difficult to calculate the uncertainty derived from the particle deformation based on the propagation law of uncertainty. This is a conservative estimate, but assuming a rectangular distribution with half the amount of particle deformation $\Delta d/2$, the standard uncertainty derived from the particle deformation, $u(\Delta d)$ can be calculated by the following equation,

$$u(\Delta d) = (\Delta d/2) \cdot \left(1/\sqrt{3}\right). \tag{6.5}$$

Table 6.4 shows the calculated values of particle deformation, Δd, corrected particle sizes, d, and their expanded uncertainties, U_{95}. Figure 6.5 shows a comparison of the nano particle size in the height direction before correction and its expanded uncertainty, the nano particle size in the height direction after correction and its expanded uncertainty, and the nominal value of the nano particle size and its expanded uncertainty. The vertical axis of the graph represents the difference in particle size from the nominal value. The particle sizes in the height direction after the particle size correction are included in the range of the expanded uncertainties of the nominal values.

Table 6.3. Material properties. Reprinted from Misumi *et al* 2018. Copyright (2018), with permission from ASPE, EUSPEN and JSPE.

Material	Surface energy	Yield point
PSL	45 $mN \cdot m^{-1}$	9 MPa
Mica	47.7 $mN \cdot m^{-1}$	

Table 6.4. The calculated particle deformation, the corrected mean particle diameter, d, and the expanded uncertainty after correction, U_{95}. Reprinted from Misumi *et al* 2018. Copyright (2018), with permission from ASPE, EUSPEN and JSPE.

Sample	PSL 30 nm	PSL 100 nm	PSL 300 nm
Calculated particle deformation, Δd	4.0 nm	3.6 nm	3.5 nm
Corrected mean particle diameter (after correction), d	27.3 nm	99.0 nm	305.5 nm
Expanded uncertainty (after correction), U_{95}	2.5 nm	2.3 nm	2.1 nm

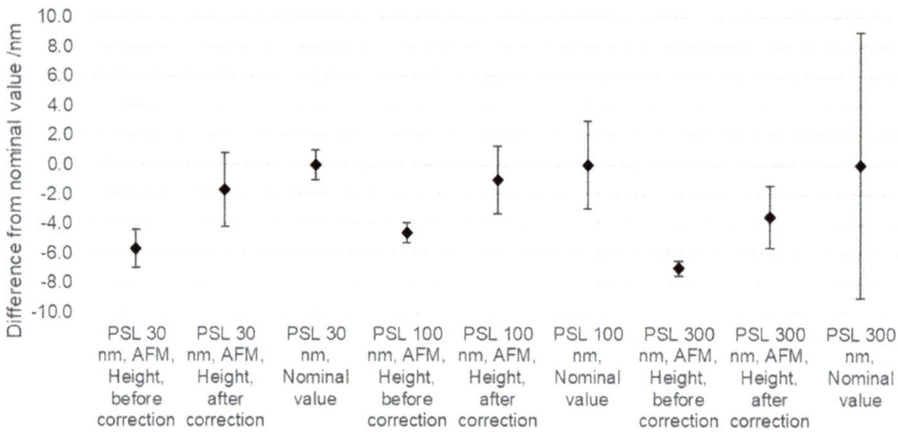

Figure 6.5. Particle sizes (heights) before and after correction and nominal values of PSL standard nano particles.

6.4 Summary

Size measurement methods of PSL nano particles using a metrological AFM were explained. The calculation method of particle deformation was also introduced.

References

Garnaes J 2011 Diameter measurements of polystyrene particles with atomic force microscopy *Meas. Sci. Technol.* **22** 094001

Lin H-L *et al* 2019 Nanoparticle characterization —supplementary comparison on nanoparticle size *Metrologia* **56** 1A

Maugis D and Pollock H M 1984 Surface forces, deformation and adherence at metal micro-contacts *Acta Metal.* **32** 1323–34

Meli F 2005 Lateral and vertical diameter measurements on polymer particles with a metrology AFM *Nanoscale Calibration Standards and Methods: Dimensional and Related Measurements in the Micro- and Nanometer Range* ed G Wilkening and L Koenders (Weinheim: Wiley-VCH Verlag) ch 27

Meli F *et al* 2012 Traceable size determination of nanoparticles, a comparison among European metrology institutes *Meas. Sci. Technol.* **23** 125005

Meli F and Küng A 2011 Realization of a large sample 3D metrology AFM with differential Jamin interferometers *Proc. of the 11th Euspen Int. Conf.—Como— (May 2011)*

Misumi I, Sugawara K, Takahata K, Takahashi K and Ehara K 2018 Size measurements of standard nanoparticles using metrological atomic force microscope and evaluation of their uncertainties *Precis. Eng.* **51** 691–701

Rimai D S, Quesnel D J and Busnaina A A 2000 The adhesion of dry particles in the nanometer to micrometer-size range *Colloids Surf.* A **165** 3–10

Wang X D, Chen B, Wang H F and Wang Z S 2011 Adhesion between submicrometer polystyrene spheres *Powder Technol.* **214** 447–50

IOP Publishing

Nanoscale Standards by Metrological AFM and Other Instruments

Ichiko Misumi

Chapter 7

Surface roughness

Surface roughness has been measured by other types of instruments, for instance, stylus and optical instruments. Recently, a metrological AFM has also been used for surface roughness measurements. Evaluation of the probe-tip shape is the key point to roughness measurement. An example of roughness measurement and uncertainty evaluation performed by the method based on ISO 19606 is provided.

7.1 Introduction

Conventionally, surface roughness has been measured by stylus instruments. Surface roughness measurement using AFMs has also been performed in recent years. In order to evaluate the surface roughness measurement results by AFMs, comparisons with other measurement methods have been performed. DFM, the Danish NMI, compared surface roughness measurements between an AFM and an optical metrology tool (Garnaes *et al* 2003). In addition, the PTB in Germany compared the surface roughness measurement with a metrological AFM and a stylus instrument (Dai *et al* 2004).

Surface roughness measurements based on the document standards are required. This is because it is difficult to judge whether surface roughness measurement results are correct unless the measurements are based on the document standard. Some document standards for surface texture have been published in the ISO 25178 series under the direct responsibility of ISO TC213 (Technical committee: Dimensional and geometrical product specifications and verification). Atomic force microscopy is considered as one of the methods for three-dimensional surface topography measurements in 3.3.15 of 'ISO 25178-6: Geometrical product specifications (GPS) —Surface texture: Areal—Part 6: Classification of methods for measuring surface texture'. However, it is stated in NOTE of 3.3.15 of ISO 25187-6 that the development of certain standards related technologies such as AFM falls under the scope of

doi:10.1088/978-0-7503-3191-3ch7

ISO TC201 (Surface science analysis), SC9 (Scanning probe microscopy). Document standards of the characteristics of measuring instruments have been published, for instance, the stylus scanning method in ISO 25178-601 and the chromatic aberration confocal microscopy in ISO 25178-602. A document standard for AFM has not yet been published in the ISO 25178 series. Therefore, according to 3.3.15 Note 1, it is necessary to confirm the standardization status of roughness measurement by AFM in ISO TC201, SC9.

Table 7.1 shows the organization of ISO TC201, SC9. ISO TC201, SC9 has SG3 (study group 3) nanoscale roughness measurement. Table 7.2 shows the standards published or being discussed by ISO/TC201/SC9. Seven document standards have been issued and two are under discussion. As shown in table 7.2, a document standard for roughness measurement by AFM has not reached the above stage yet.

There is a significant demand from industries for surface roughness measurements based on document standards. The currently available document standard for roughness measurement by AFM is 'ISO 19606 Fine ceramics (advanced ceramics, advanced technical ceramics)—Test method for surface roughness of fine ceramic films by atomic force microscopy'. This document standard was established mainly by experts on fine ceramics who are AFM users, not AFM experts. The reference standards are ISO 4287 (1997), ISO 4288 (1996), ISO 18115-2 (2013), ISO 11039 (2012), ISO 11952 (2019), ISO 25178-2 (2012), and ISO 19606 (2017). These document standards were established by experts on surface roughness measurement and AFM. An example of surface roughness measurement based on ISO 19606 (Misumi *et al* 2015, 2019, 2020) is given in section 7.2.

7.2 Measurement conditions

7.2.1 Instruments

The *X*- and *Y*-axis of an AFM scanner are calibrated using a calibrated one-dimensional grating or two-dimensional grating, and the *Z*-axis of an AFM scanner

Table 7.1. Organization of ISO/TC201 (surface science analysis)/SC9 (scanning probe microscope). SC9 has a study group on 'nanoscale roughness measurement'.

Number	Organization
CAG	Chairman's advisory group
SG1	Effects of temperature and humidity on dimensional atomic force microscopy measurements
SG2	Use of NSOM/SNOM
SG3	**Nanoscale roughness measurement**
SG4	Use of STM
WG3	Basic dimensional calibration of SPMs
WG4	Application-oriented dimensional SPM calibrations
WG5	Calibration of probes
WG6	Use of ESPM
WG7	Mechanical properties of materials

Table 7.2. Published ISO standards and ISO standards under development under the direct responsibility of ISO/TC201/SC9. No document standards concerning nanoscale roughness measurement have been published yet.

Number	Title
ISO 11039	Measurement of drift rate
ISO 11775	Determination of cantilever nominal spring constant
ISO 11952	Determination of geometric quantities using SPM: Calibration of measuring systems
ISO 13083	Standards on the definition of spatial resolution of electrical scanning probe microscopes (ESPMs) such as SSRM and SCM for 2D-dopant imaging and other purposes
ISO 13095	Procedure for *in situ* characterization of AFM probes' shank profile used for nanostructure measurement
ISO 21222	Guideline for the determination of elastic moduli for compliant materials using an atomic force microscope
ISO 27911	Definition and calibration of the lateral resolution of a near-field optical microscope
ISO/CD 23683 [Under development]	Guideline for experimental quantification of carrier concentration in semiconductor devices by using electric scanning probe microscopy
ISO/WD 23729 [Under development]	Guideline for restoration procedure for atomic force microscopy images dilated by finite probe size

is calibrated using a calibrated step height standard when a commercially available AFM is used. Surface roughness measurement that is traceable to the SI can be realized when using a metrological AFM in which the laser interferometer is installed on the *XYZ* axes. Therefore, no calibration procedures specified in ISO 19606 for the *X*-, *Y*-, and *Z*-axes of AFM scanners using calibrated nano dimensional standards are required.

7.2.2 Method to determine criteria for probe-tip error

Figure 7.1 shows the probe-tip evaluation sequence. *Ra* and *RSm* are the arithmetic mean roughness and the average width of the roughness profile element, respectively. Firstly, a preliminary roughness measurement of a specimen is performed to obtain preliminary *Ra* and *RSm*. Preliminary *Ra* and *RSm* are necessary when judging whether the probe-tip can be used in a surface roughness measurement or not. Next, a part of the probe-tip diameter evaluation standard plate is measured. Spikes are formed on the surface of the probe-tip diameter evaluation standard plate. A cross-section of the spike profile 10 nm below the top in the *Z* direction is obtained and *D* is calculated as the probe cross-section diameter based on ISO 19606 '10.3 Probe-tip diameter evaluation' and 'Annex A'. The desired value for *D*, the cross-sectional diameter of the AFM probe-tip for measuring surface roughness with sufficient accuracy, is expressed using the following model. The model assumes that

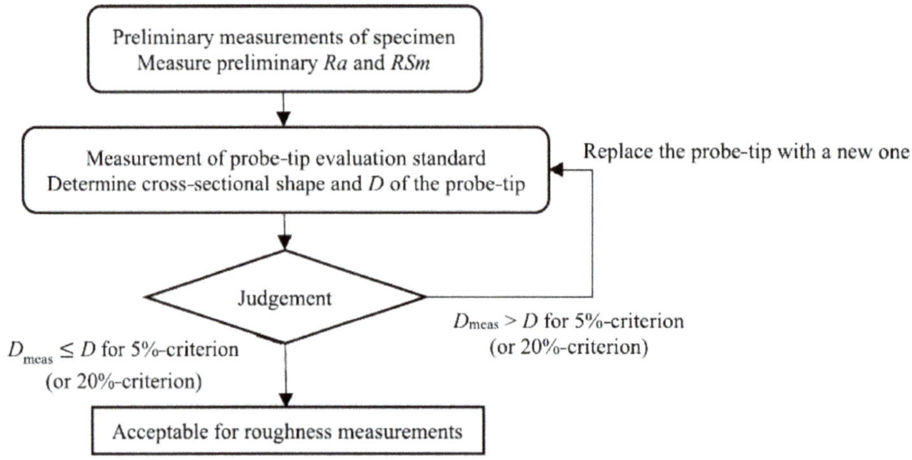

Figure 7.1. Probe-tip evaluation sequence (Misumi *et al* 2020).

the sample surface profile is a sine wave with a period of *RSm* given by equation (7.1).

$$f(x) = \frac{\pi Ra}{2} \cdot \cos\left(\frac{2\pi x}{RSm}\right). \tag{7.1}$$

In this case, *Ra* is expressed as follows

$$Ra = \frac{Rz}{\pi}, \tag{7.2}$$

where *Rz* is the maximum height of roughness profile.

Equation (7.2) gives the values between the triangular and rectangular wavefront. This is often close to the actual value of the specimen. The probe-tip is assumed to be a sphere with a radius of curvature *r*, which is desirable if it can reach the bottom parts of the roughness profile. The probe-tip whose radius is larger than the maximum radius ($r = r_{max}$) of the inscribed circle of the minimum part of the roughness profile cannot reach the minimum part of the roughness profile and is not suitable for roughness measurement. Probe-tips with *r* below r_{max} are considered acceptable. Based on the above model, r_{max} is expressed by the following equation,

$$r_{max} = \frac{RSm^2}{2\pi^3 Ra}. \tag{7.3}$$

Therefore, D_{max}, which is the value of *D* related to r_{max}, is expressed by the following equation,

$$D_{max} = 2\sqrt{20r_{max} - 100}. \tag{7.4}$$

However, for sharp probe-tips with $r < 10$ nm, $D_{max} = 2\ r_{max}$ should be used. *Ra* is a constant, its value is the approximate value obtained in the first step of figure 7.1, and *RSm* is a variable in the range 0.04 μm to 2.5 μm. Figure 7.2 shows the *D-RSm* map described based on appendix B of ISO 19606. The preliminary *Ra* and *RSm* are 5 nm and 100 nm, respectively. The evaluation result when $D = 15$ nm is 'Evaluation result 1' in figure 7.2, the evaluation result when $D = 30$ nm is 'Evaluation result 2' in figure 7.2, and the evaluation result when $D = 60$ nm is 'Evaluation result 3' in figure 7.2, respectively. When D is less than $D_{max}/2$ [5% standard] as shown in Evaluation result 1, the *Ra* measurement error is expected to be less than 5%. If D is between $D_{max}/2$ and D_{max} as shown in Evaluation result 2 [reference 20%], the measurement error of *Ra* is expected to be less than 20%. Evaluation result 3 indicates that D is too large and the probe-tip is not suitable for roughness measurement.

7.2.3 Standard plate used for probe-tip diameter evaluation

Table 7.3 shows the specifications of the probe-tip diameter evaluation standard plates. TGT1 was used in the previous studies (Misumi *et al* 2015, 2019). PA01 was used in a recent study (Misumi *et al* 2020). Figures 7.3(a)–(c) show a schematic drawing of the probe-tip diameter evaluation standard plate used in the previous studies, a thermal field emission scanning electron microscope (FE-SEM) (HITACHI S4300-SE, acceleration voltage: 2.0 kV) image of a top view of the standard plate, and an SEM image of a bird's-eye view of the standard plate,

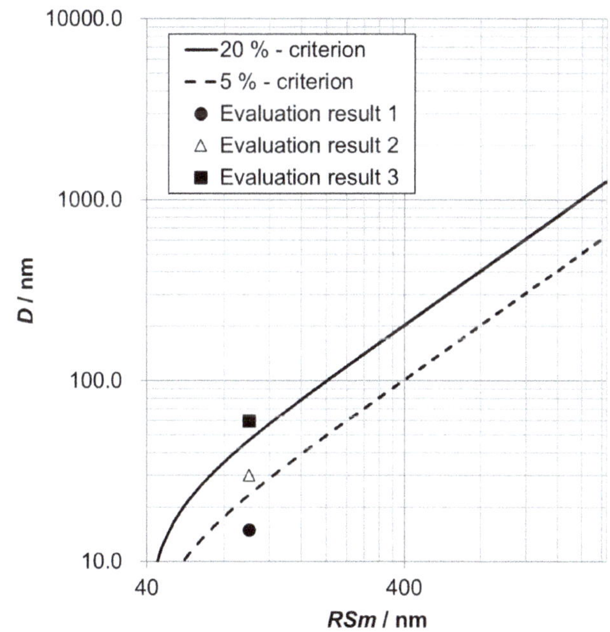

Figure 7.2. A *D-RSm* map for the probe-tip error evaluation to measure the roughness of a specimen ($Ra = 5$ nm, $RSm = 100$ nm). When $D = 15$ nm, the probe-tip error evaluation result is 'Evaluation result 1'. When $D = 30$ nm, the probe-tip error evaluation result is 'Evaluation result 2'. When $D = 60$ nm, the probe-tip error evaluation result is 'Evaluation result 3'.

Table 7.3. Specification of the standard plate used for probe-tip diameter evaluation (Misumi *et al* 2020).

Manufacturer	Product number	Distance between any two nearest neighboring spikes (nm)	Spike base (nm)	Spike height (nm)	Spike angle (degrees)	Spike curvature radius (nm)	Reference
NT-MDT Spectrum Instruments	TGT1	2120		300–500	50	10	Misumi *et al* (2015, 2019)
MICROMASCH®	PA01	(100–200)	50–100	50–150		Less than 5 (minimum)	Misumi *et al* (2020)

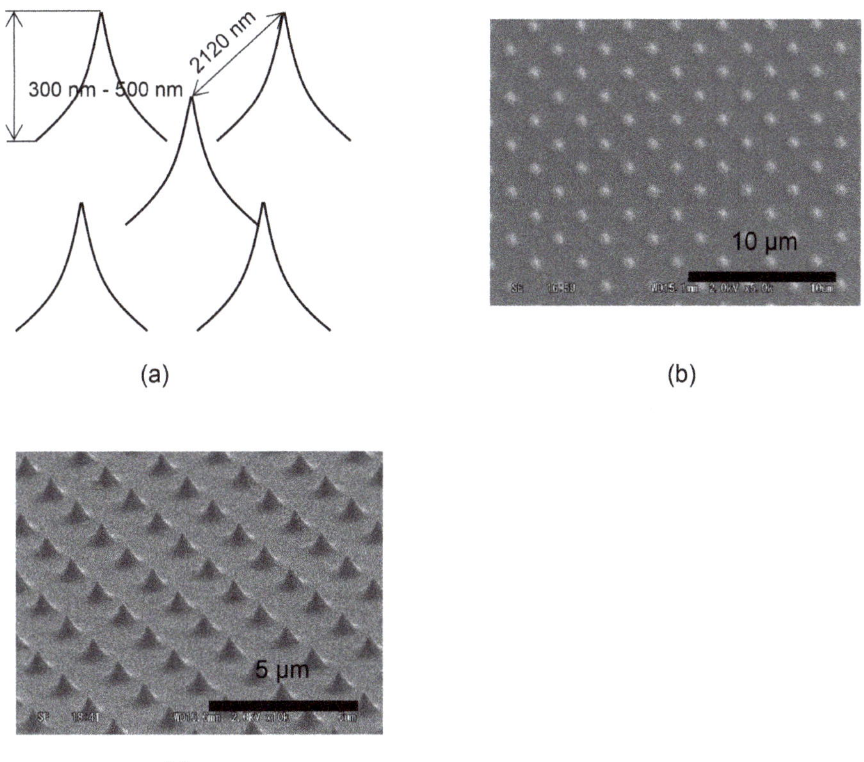

Figure 7.3. (a) Schematic drawing of a probe-tip diameter evaluation standard plate used in references (Misumi *et al* 2015, 2019). (b) Thermal field emission scanning electron microscope (FE-SEM) (HITACHI S4300-SE, acceleration voltage: 2.0 kV) image of a top view of the standard plate. (c) SEM image of a bird's-eye view of the standard plate (Misumi *et al* 2020).

respectively. The standard plate has many needle-shaped spikes and has the advantage of being intuitive and easy to use because it has a shape similar to the knife edge used in conventional stylus-type roughness measuring instruments. However, previous studies have identified some drawbacks. (1) Needle spikes are too high compared to the sample surface roughness (generally in the 300 nm to 500 nm range). When evaluating the probe-tip diameter before measuring the specimen roughness, the probe-tip may be damaged. Since it is impossible to know the coordinates of a particular needle spike in advance, AFM measurement that captures the overall shape of the needle spike is necessary. At this time, the AFM probe-tip needs to scan the slope of the spike. AFM probe control along the vertical axis tends to be unstable at steep slopes due to interference between the probe-tip (which has a finite cone angle) and the spike. As a result, tip wear can be significant. Moreover, spike measurements are not always completed at one time due to broken needle spikes. Figure 7.4 shows a metrological AFM image of a standard plate used to assess the probe-tip diameter. Figure 7.4(a) shows an example of a successful

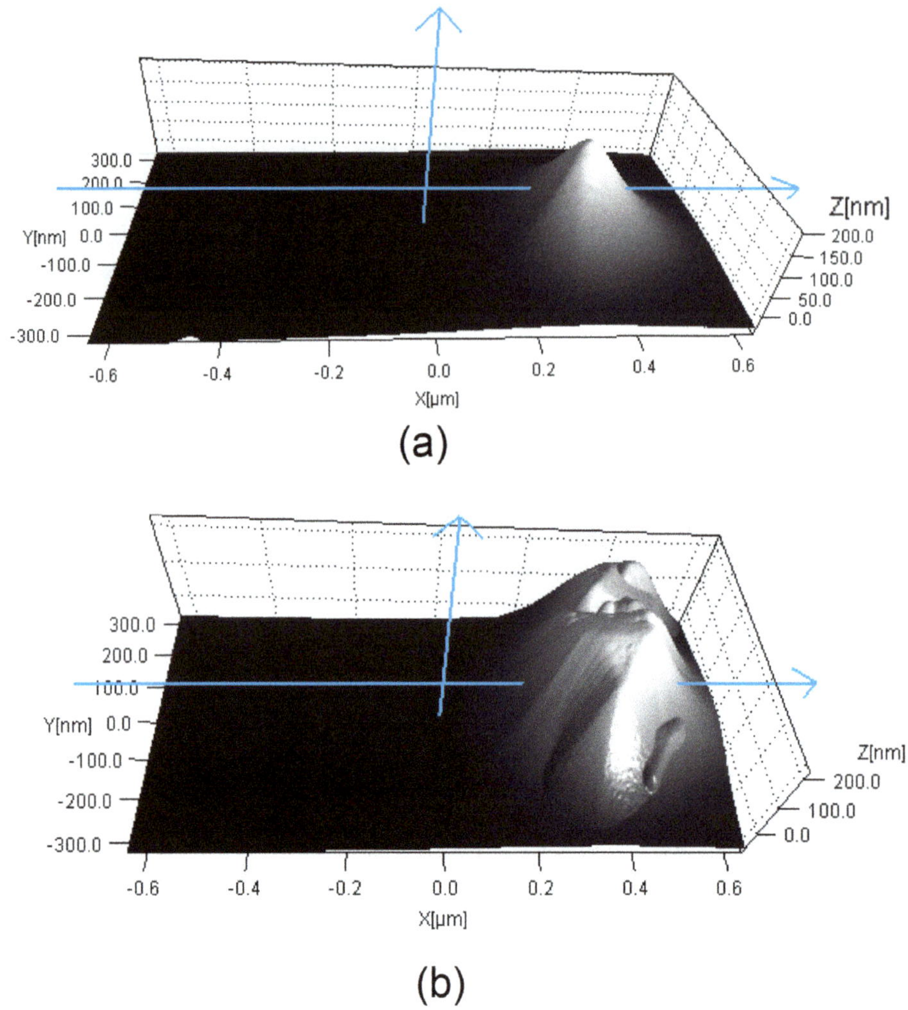

Figure 7.4. Metrological AFM images of the standard plate (TGT1) used for probe-tip diameter evaluation. (a) Example of a successful approach of a probe-tip to the standard plate. (b) Example of a failure to approach the same probe-tip to the standard plate (Misumi *et al* 2020).

probe-tip approach to a standard plate, and figure 7.4(b) shows an example of failure to approach the same probe-tip to a standard plate. (2) The distance between spikes is about 2.12 μm, which is too long. Positioning the standard plate requires multiple AFM images that are time consuming. (3) The radius of curvature of the spike is about 10 nm, which is not small enough compared to the AFM probe-tip size. If the spike radius of curvature is small enough, the probe-tip diameter can be overestimated.

Figures 7.5(a)–(c) show a schematic drawing of the top view of the standard plate for evaluating the probe-tip diameter used in recent research (Misumi *et al* 2020), a

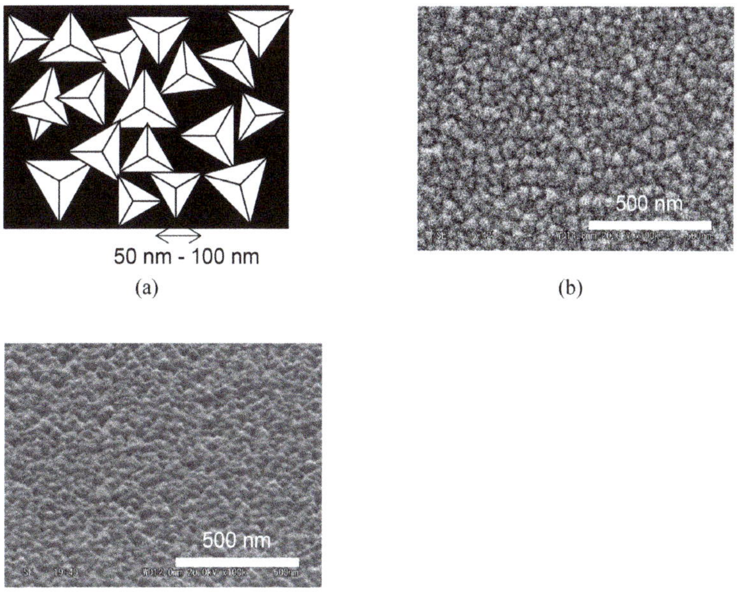

(a) (b) (c)

Figure 7.5. (a) Schematic drawing of the top view of the probe-tip diameter evaluation standard plate. Thermal FE-SEM (acceleration voltage: 20.0 kV) image of the top view of the standard plate. (c) SEM image of a bird's-eye view of the standard plate (Misumi *et al* 2020).

thermal FE-SEM (accelerating voltage: 20.0 kV) image of a top view of the standard plate, and an SEM image of the bird's-eye view of the standard plate, respectively. This standard plate consists of sharp pyramidal nanostructures. The pyramids are triangular with a base of 50 nm to 100 nm in length and their height ranges from 50 nm to 150 nm. The radius of curvature of the sharpest tips is shorter than 5 nm.

7.2.4 Specimens and cantilever probes

Table 7.4 shows the specifications of the specimens for roughness measurements. Standard silicon cantilever probes (OMCL-AC160TS, Olympus Corporation) were used. The spring constant is approximately 42 N · m^{-1} and the resonance frequency is approximately 300 kHz.

7.2.5 Roughness measurements

Table 7.5 shows the roughness measurement conditions for each specimen. The reason why the number of pixels is 512(Y) × 16(X) is as follows. The number of pixels in the fast scan direction and the slow scan direction are selected from 256 pixels to 512 pixels, and 10 pixels to 512 pixels, respectively, based on the '11 sample roughness measurement' of ISO 19606. Furthermore, in the final step of roughness measurement, the S_a (arithmetic mean height of the scale-limited surface) is characterized by 512(Y) × 512(X) pixels at the fifth measurement point of all eight

Table 7.4. Specification of the specimens (Misumi *et al* 2020).

Specimen No.	Specimen	Manufacturer	Product number
1	Silicon wafer	Shin-Etsu Chemical Co. Ltd.	CZ-P <100>
2	Flat aluminum total reflection mirror	SURUGA SEIKI CO., Ltd.	S01-20-3T
3	Indium tin oxide (ITO) thin film	GEOMATEC CO., Ltd.	0007
4	Indium tin oxide (ITO) thin film	GEOMATEC CO., Ltd.	0005
5	TipCheck sample	Aurora NanoDevices Inc.	TIP001
6	Gauge block	Mitutoyo Corporation	613614-02
7	Roughness standard	SiMETRICS GmbH	RnS 40
8	Roughness standard	SiMETRICS GmbH	ARS f1

Table 7.5. Roughness measurement conditions of each specimen (Misumi *et al* 2020).

Specimen No.	Evaluation length (μm × μm)	Scanning speed(nm s^{-1})	Number of picture elements	Number of measuring points	Repetition number at each measuring point
1	1 × 1	100	512 × 16	5	2
2	2 × 2	400	512 × 16	5	2
3	2 × 2	400	512 × 16	5	2
4	5 × 5	500	512 × 16	5	2
5	5 × 5	500	512 × 16	5	2
6	10 × 10	1000	512 × 16	5	2
7	20 × 20	1200	512 × 16	5	2
8	20 × 20	1200	512 × 16	5	2

specimens. All S_a were within the uncertainty of all *Ra*. Furthermore, as shown in figure 7(a) of the previous study (Misumi *et al* 2015), when the measurement was repeated with 512 × 512 pixels, the probe-tip was worn, the roughness value decreased, and the uncertainty increased. Therefore, the number of pixels of 512 (Y) × 16(X) is suitable. One probe is used per specimen. The roughness was measured in the order of Specimens 1–8. The measurement order of the five measurement points in a specimen was randomly selected. Two repeated measurements were continuously performed at each measurement point. This is because NMIJ's metrological AFM scanner cannot be repositioned at the same measurement point. *Ra*, *RSm*, and *D* were calculated by combining the roughness analysis module with the particle analysis module of the Scanning Probe Image Processor (from Image Metrology A/S). The λc filter is set to one fifth of each scanning range.

7.2.6 Uncertainty evaluation

The main sources of uncertainty are the same as one-dimensional gratings and step height standards. The sources are non-uniformity of surface roughness of a specimen, repeatability of measurement, interferometer non-linearity, interferometer resolution, Abbe error, correction of refractive index in air, the change of the optical path difference in interferometer, and the stability of laser frequency. The source of uncertainty peculiar to roughness measurement is the uncertainty derived from probe-tip error evaluation. The standard uncertainty caused by the probe-tip error evaluation is expressed as follows.

$$\begin{aligned} u_{\text{tip}} &= 0.2Ra\text{: } D \leqslant D_{\max}, \\ u_{\text{tip}} &= 0.05Ra\text{: } D \leqslant D_{\max}/2. \end{aligned} \tag{7.5}$$

7.3 Measurement results and discussion

7.3.1 Evaluation of probe-tip diameter

Figure 7.6(a) shows a 3D rendered image of the metrological AFM image of the standard plate captured before measuring the surface roughness of Specimen 2. The scan area is 320 nm (*Y*) × 320 nm (*X*). The number of pixels is set to 512 pixels (*Y*) × 512 pixels (*X*) based on '10.3 Probe-tip diameter evaluation' of ISO 19606. Figure 7.6(b) is the gradient norm image of figure 7.6(a). Figure 7.6(c) is a top view of figure 7.6(a). The numbers in figure 7.6(c) show the locations of the spikes, the cross-sections of the spikes are profiled 10 nm below the tops in the *Z* direction and *D* are calculated. Figure 7.6(d) shows an example of a cross-sectional area (blue) 10 nm below in the *Z* direction from the top. Once each *D* is calculated, the area of interest (AOI) is limited to contain only one spike, as shown in figure 7.6(d). Table 7.6 shows the calculation results of *D*. The range of the calculation results is from 30.4 nm at position 2 to 37.0 nm at position 8. The minimum value of 30.4 nm is selected as *D* for specimen 2. This probe-tip diameter evaluation standard plate eliminates the need to align the AFM probe-tip to a specific spike and makes it possible to measure many spikes at once. This can reduce tip wear. The minimum value (taken from the sharpest spike on the standard plate) can be selected to estimate the optimal *D*. In the other seven specimens, *D* ranged from 21.2 nm to 38.2 nm.

Figure 7.7 shows the *D-RSm* map of the error evaluation of the probe-tip in the roughness measurement of Specimen 2 and the evaluation results. More information on the *D-RSm* map is given in section 7.2.2. The preliminary *Ra* and preliminary *RSm* are 1.4 nm and 80.0 nm, respectively. The solid and dotted lines represent the 20% and 5% criteria, respectively, and the solid circle is below the dotted line.

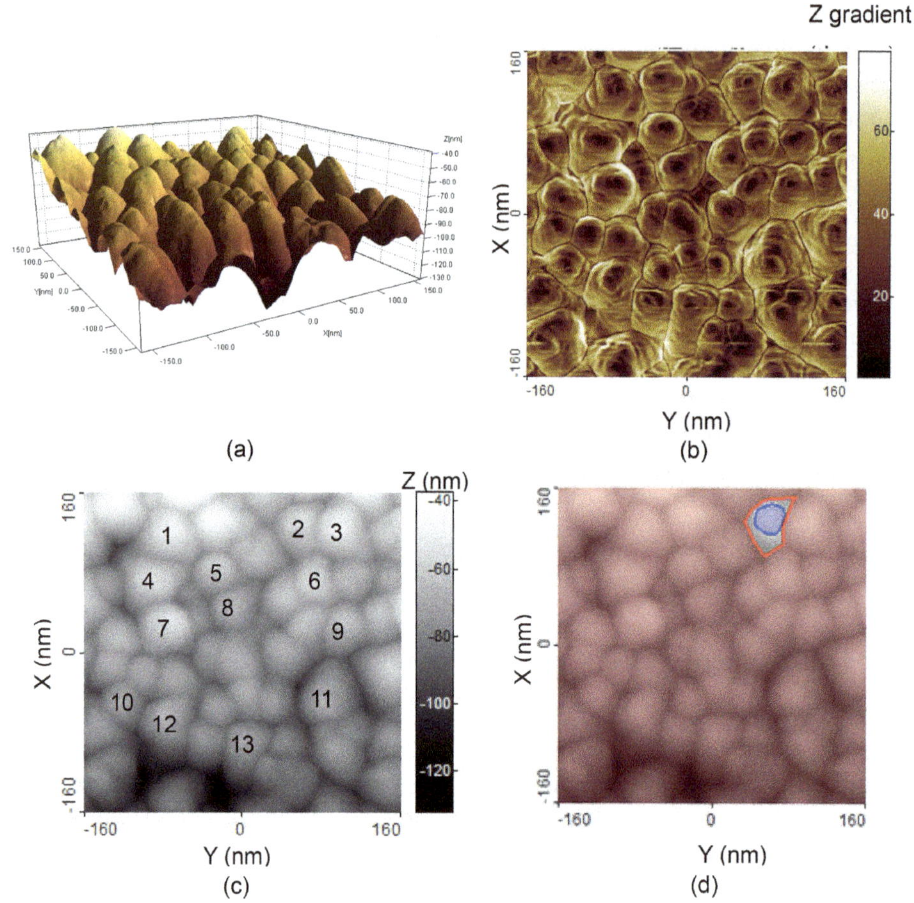

Figure 7.6. (a) A 3D rendering image of a metrological AFM image of the standard plate used for probe-tip diameter evaluation. (b) The *Z* gradient norm image of (a). (c) Top view of (a): The numbers indicate spike positions where the cross-sections of the spikes profiled at 10 nm below their tops in the *Z* direction and the *D* were calculated. (d) An example of cross-sectional area (blue color) at 10 nm below its top in *Z* direction (Misumi *et al* 2020). Copyright IOP Publishing Ltd. All rights reserved.

This shows that the probe-tip can make roughness measurements with less than 5% error. With respect to the other seven specimens, the error evaluation results of the probe-tips showed that the surface roughness can also be measured with an error of less than 5%.

7.3.2 Roughness measurement results

Table 7.7 shows the *Ra* measurement result and expanded uncertainty of each specimen. The major sources of uncertainty and their contribution to the combined standard uncertainty are shown in table 7.8. The interferometer non-linearity and the interferometer resolution are constant regardless of *Ra*. The uncertainty derived from the probe-tip error evaluation increases as *Ra* increases.

Table 7.6. Calculation results of probe-tip diameters, *D* (Misumi *et al* 2020). Copyright IOP Publishing Ltd. All rights reserved.

Spike position No.	Probe-tip diameter, *D* (nm)	Area (nm^2)	Minimum *Z* (nm)	Maximum *Z* (nm)
1	35.3	978.04	−58.02	−48.06
2	30.4	727.45	−63.04	−53.08
3	34.2	920.78	−57.71	−47.75
4	36.6	1052.5	−65.09	−55.13
5	33.4	875.68	−69.15	−59.18
6	33.4	873.72	−64.78	−54.82
7	35.5	986.66	−64.67	−54.70
8	37.0	1075.7	−76.14	−66.17
9	34.3	923.92	−71.66	−61.69
10	31.4	776.47	−77.61	−67.64
11	35.5	989.02	−76.18	−66.21
12	32.9	849.02	−74.90	−64.94
13	33.6	888.62	−78.77	−68.72

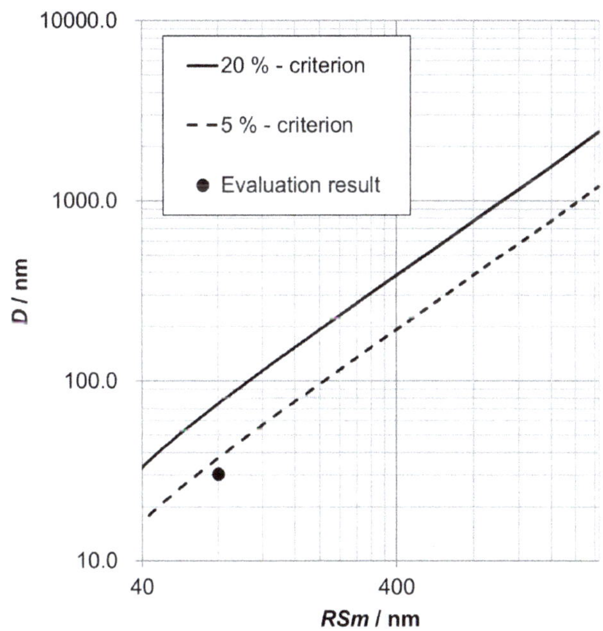

Figure 7.7. A *D-RSm* map for the probe-tip error evaluation to measure roughness of Specimen 2 (Misumi *et al* 2020). Copyright IOP Publishing Ltd. All rights reserved.

Table 7.7. Roughness measurement results of each specimen (Misumi *et al* 2020).

Specimen No.	Measurement result, *Ra* (nm)	Expanded uncertainty, U_{95} (nm)
1	0.15	0.14
2	1.3	0.2
3	1.5	0.2
4	4.4	0.4
5	8.4	0.5
6	11.7	2.3
7	63.7	4.2
8	103.8	9.1

Table 7.8. Major sources of uncertainty in surface roughness measurements of all specimens and their contributions to the combined standard uncertainty (Misumi *et al* 2020).

Major source of uncertainty	Contribution to the combined standard uncertainty, *u* (nm)								
	Specimen No.	1	2	3	4	5	6	7	8
Interferometer non-linearity		0.07	0.07	0.07	0.07	0.07	0.07	0.07	0.07
Probe-tip error evaluation		0.004	0.04	0.04	0.13	0.24	0.34	1.8	3.0
Non-uniformity of surface roughness of a specimen		Negligible	0.04	0.02	0.11	0.09	0.82	0.7	2.9
Interferometer resolution		0.02	0.02	0.02	0.02	0.02	0.02	0.02	0.02
Repeatability of measurement		Negligible	Negligible	Negligible	0.01	Negligible	0.03	0.03	0.05

7.4 Summary

The method of measuring the arithmetic mean roughness *Ra* based on ISO 19606 and its uncertainty evaluation method were explained. This method can also be applied to S_a measurements when a more suitable probe-tip can be selected. When S_a can be measured by a metrological AFM and the uncertainty can be evaluated, it is possible to compare measurement results between a metrological AFM and optical metrology instruments.

References

Dai G, Jung L, Pohlenz F, Danzebrink H U, Krueger-Sehm R, Hasche K and Wilkening G 2004 Measurement of micro-roughness using a metrological large range scanning force microscope *Meas. Sci. Technol.* **15** 2039

Garnaes J, Kofod N, Kuehle A, Nielsen C, Dirscherl K and Blunt L 2003 Calibration of step heights and roughness measurements with atomic force microscopes *Precis. Eng* **27** 91

ISO 4287 1997 Geometrical Product Specifications (GPS)—Surface Texture: Profile Method—Terms, Definitions and Surface Texture Parameters

ISO 4288 1996 Geometrical Product Specifications (GPS)—Surface Texture: Profile Method—Rules and Procedures for the Assessment of Surface Texture

ISO 11039 2012 Surface Chemical Analysis—Scanning-probe Microscopy—Measurement of Drift Rate

ISO 11952 2019 Surface Chemical Analysis—Scanning-probe Microscopy—Determination of Geometric Quantities using SPM: Calibration of Measuring Systems

ISO 18115-2 2013 Surface Chemical Analysis—Vocabulary—Part 2: Terms Used in Scanning-probe Microscopy

ISO 25178-2 2012 Geometrical Product Specifications (GPS)—Surface Texture: Areal—Part 2: Terms, Definitions and Surface Texture Parameters

ISO 19606 2017 Fine Ceramics (Advanced Ceramics, Advanced Technical Ceramics)—Test Method for Surface Roughness of Fine Ceramic Films by Atomic Force Microscopy

Misumi I, Naoi K, Sugawara K and Gonda S 2015 Profile surface roughness measurement using metrological atomic force microscope and uncertainty evaluation *Measurement* **73** 295

Misumi I, Sugawara K, Kizu R, Hirai A and Gonda S 2019 Extension of the range of profile surface roughness measurements using metrological atomic force microscope *Precis. Eng.* **56** 321

Misumi I, Sugawara K, Kizu R, Hirai A and Gonda S 2020 A standard used for probe-tip diameter evaluation in surface roughness measurements using metrological atomic force microscope *Meas. Sci. Technol.* **31** 094005

Ichiko Misumi

Chapter 8

Secondary realization of the SI meter using silicon lattice parameters

Firstly, the status of nano dimensional standards is summarized. Next, the secondary realization of an SI meter using silicon lattice parameters is introduced. Finally, the impact on the industry due to the release of CCL-GD-MeP-1 is described.

8.1 The status of nano dimensional standards

Traceability has been mainly ensured through nano dimensional standards calibrated to being SI traceable in industrial fields. National metrology institutes (NMIs) have mainly developed AFMs equipped with laser interferometers (metrological AFMs) to calibrate nano dimensional standards. Metrological AFMs have ensured traceability to SI by calibrating the frequencies of the light sources of the laser interferometers. However, the miniaturization of nano dimensional standards is progressing rapidly, and the uncertainty in the calibration of the nano dimensional standards is not sufficiently small due to the uncertainty derived from the metrological AFMs such as the periodic error of the laser interferometer (Misumi *et al* 2019). This limit has become apparent in the fact that only laser frequencies have been recommended as *Mise en pratique* (MeP) for the definition of the meter in the SI.

8.2 Secondary realization of the SI meter using silicon lattice parameters

The International Committee for Weights and Measures (CIPM), Consultative Committee for Length (CCL) recommended the value of Si lattice parameters as MeP for the definition of the meter in the SI in 2019 (Dixson *et al* 2019, Koenders

doi:10.1088/978-0-7503-3191-3ch8

et al 2019, Yacoot *et al* 2019, 2020). Three guidelines for secondary realizations were also proposed in 2019 as follows.

CCL-GD-MeP-1: Secondary realization of the SI meter using silicon lattice parameter and X-ray interferometry for nanometer and sub-nanometer scale applications in dimensional nanometrology

CCL-GD-MeP-2: Secondary realization of SI meter using silicon lattice and transmission electron microscopy for dimensional nanometrology

CCL-GD-MeP-3: Secondary realization of SI meter using height of monoatomic steps of crystalline silicon surfaces

The technical backgrounds of the three guidelines are introduced as follows.

8.2.1 CCL-GD-MeP-1

The X-ray interferometry technique combined with knowledge of silicon lattice spacing provides a metrological standard with sufficiently small periodicity. The lattice constant of silicon has been determined on bulk silicon crystals with relative uncertainties up to 10^{-8} using optical and X-ray interferometry (Mohr *et al* 2012). The value of Si {220} lattice spacing d_{220} is 192.015 5714 × 10^{-12} m and its standard uncertainty is 0.000 0032 × 10^{-12} m. The fringe spacing of the X-ray interferometer is purely based on the lattice constant of the crystal in which the X-ray is diffracted and is independent of the wavelength of the X-ray used. Since this technique is based on counting the atoms in the crystal, there is no significant non-linearity when using X-ray interferometry. The lattice constant of silicon sets the effective periodicity to 0.192 nm when X-rays are diffracted from the d_{220} plane. Integer-order subdivision with low lattice spacing is possible with proper X-ray interferometer configuration, which can reduce resolution to a few picometers with slight non-linearity at this level (Yacoot and Kuetgens 2012).

NPL, PTB, and formerly IMGC (now INRIM, NMI in Italy) have developed the Combined Optical and X-ray Interferometer (COXI) to use the Si d_{220} lattice spacing as a reference standard for dimensional measurements using X-ray interferometry (Basile *et al* 2000). This establishes traceability to the meter via both the laser frequency of the He–Ne laser and the lattice constant of silicon, which was previously measured using X-ray interferometry. Long-distance measurements (up to ±1 mm) were made using an optical interferometer, and short-range, high-precision measurements were made using an X-ray interferometer that operates on the same principle as the vernier scale. This has eliminated the need for an optical fringe division. Subsequent collaboration between NPL and PTB has led to the evaluation of several displacement measuring transducers and the use of X-ray interferometers as the positioning stage for scanning probe microscopes (Yacoot *et al* 2001). Sub-X-ray fringe positioning capability has also been demonstrated (Yacoot and Kuetgens 2012).

Silicon is the preferred choice for X-ray interferometer construction, not only because its lattice constants are widely known, but also because it is available as a pure defect-free crystal in the form of rods with a particular crystal orientation. Figure 8.1 shows a schematic drawing of an X-ray interferometer along with a path

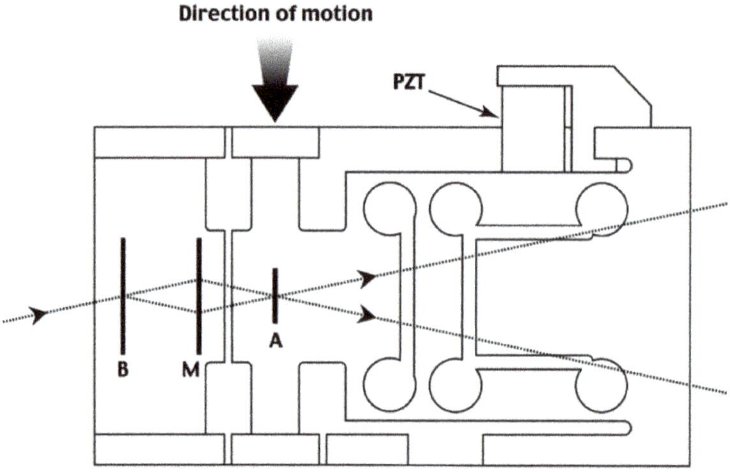

Figure 8.1. Schematic drawing of an X-ray interferometer along with a path tracked by the X-ray (Yacoot and Kuetgens 2012). Copyright IOP Publishing Ltd. All rights reserved.

Figure 8.2. Photograph of an X-ray interferometer (Yacoot and Kuetgens 2012). Copyright IOP Publishing Ltd. All rights reserved.

tracked by X-rays (Yacoot and Kuetgens 2012). Figure 8.2 shows a photograph of an X-ray interferometer (Yacoot and Kuetgens 2012).

The material is scraped from the top of the original silicon block, leaving three equally spaced thin plates, usually beam splitters (B), mirrors (M), and analyzers (A), typically hundreds of micrometers thick. The thin plate surface is usually (220) oriented perpendicular to the crystallographic plane from which it can diffract X-rays. A flexure stage is machined around the third thin plate (A, analyzer) so that the third thin plate is displaced when a force parallel to the thin plate surface is applied. During use, the interferometer is aligned so that collimated X-rays enter the beam splitter thin plate (B) at a Bragg angle with respect to the diffraction plane and are diffracted from the first thin plate (B). Two diffracted beams incident on the second thin plate (M) are generated, from which two more pairs of diffracted beams appear. The inward beams from each pair are recombined at the third thin plate (A). The combination of these two beams results in an interference pattern whose periodicity is given by the lattice parameters of the thin plate on which the X-rays are diffracted, i.e. the fringe pattern is independent of the wavelength of the X-rays used. The lattice constant of the (220) plane is approximately 0.192 nm. The third thin plane (A) is used to generate a moiré fringe pattern between the X-ray beam and the atomic plane in the crystal. As a result, when the third thin plate is displaced by a distance equal to the lattice spacing of the diffraction plane, the intensity of the X-ray beam transmitted through the third thin plate repeats maximum and minimum. By measuring the intensity of the X-ray signal when the third thin plate is displaced, the displacement of the deflection stage can be measured with respect to the lattice spacing of silicon. The range of deflection of the interferometer is a few micrometers. The stage translates using a piezo actuator. If the pitching of the stage becomes too large, the fringe contrast will decrease. The design tolerance of the flexure stage and the position of the piezo have an allowable angle error of about 10^{-8} radians. For the X-ray interferometer to be useful, the displacement of the stage must be interfaced with the outside. There are optical mirrors on the side of the X-ray interferometer, one of which is moved by a translational stage. In addition, the X-ray interferometer has a fixed optical mirror. Any optical interferometer to be evaluated can be connected to these moving and fixed mirrors. Alternatively, any bulk object to be translated can be placed directly above the third thin plate placed on the two moving optical mirrors. X-ray interferometers are capable of producing very accurate displacements and, like any precision motion system, essentially require translational capabilities with sub arc second angle errors. The Abbe cosine error is minimized when the sensor is interfaced to the system.

The X-ray source used with an X-ray interferometer is a copper Kα source with a collimating optic that is typically capable of producing a beam incident on the first thin plane, typically with an arc divergence of several minutes or longer (wavelength 0.154 nm). The shape of the beam is typically 1 mm wide and a few mm high.

The X-ray interferometer was mounted on a 500 mm × 300 mm invar breadboard to provide a larger platform for the X-ray interferometers that can accommodate optical interferometers of various sizes. This facilitated the rotation of the X-ray interferometer alignment with respect to the Bragg angle. Figure 8.3 shows the entire system.

Servo positioning of the X-ray interferometer is achieved by using the signal output of the interferometer to lock the position of the X-ray interferometer to the X-ray fringe in the center of the linear region of the fringe. By increasing the voltage applied to the PZT faster than the servo system responds, the X-ray interferometer can be moved to the next point in the center of a linear region of the same slope. The X-ray interferometer can move continuously in quantized steps corresponding to one X-ray fringe. With a digital control system, servo control between half fringes is a relatively easy procedure, as there are two linear regions with opposite gradients

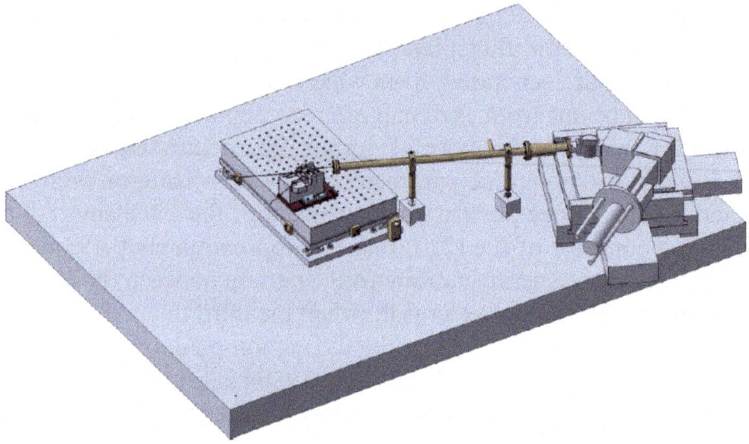

Figure 8.3. Schematic drawing of the X-ray interferometer platform with collimating optics (Yacoot and Kuetgens 2012). Copyright IOP Publishing Ltd. All rights reserved.

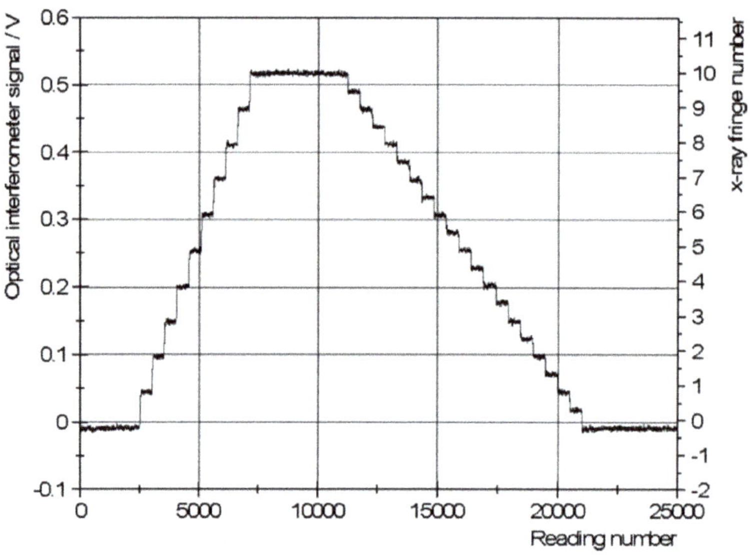

Figure 8.4. Optical interferometer signal when the X-ray interferometer moves 10 single fringe steps (10 × 192 pm) in one direction and 20 half fringe steps (20 × 96 pm) in the opposite direction (Yacoot and Kuetgens 2012). Copyright IOP Publishing Ltd. All rights reserved.

for each fringe. The step resolution quickly increases to $d_{220}/2 = 96$ pm. Figure 8.4 shows the output signals of the optical interferometer when the X-ray interferometer moves 10 single fringe steps (10 × 192 pm) in one direction and 20 half fringe steps (20 × 96 pm) in the opposite direction.

Temperature stability is very important for the X-ray interferometer. The coefficient of thermal expansion of silicon near 20°C is approximately 2.57×10^{-6} K^{-1} (Watanabe *et al* 2004). Any temperature gradient across the thin plates of the

X-ray interferometer results in a change in lattice constant and thus a reduction in fringe contrast, which results in a reduction in the available signal. The temperature non-uniformity across the thin plates should be lower than 10 mK.

Silicon single crystals used in the manufacture of X-ray interferometers should be grown ultra-pure, non-doped and dislocation-free by a float zone method with a carbon and oxygen content of less than 5×10^{15} cm^{-3}. The homogeneity of the lattice can be examined using double crystal X-ray topography, and the crystals used can be compared with those for which the lattice constant is known.

8.2.2 CCL-GD-MeP-2

Techniques such as X-ray scattering can provide a link to the Si lattice spacing for a particular measurement (particularly film thickness) defined over a wide sampling area. However, various types of transmission electron microscope (TEM) appear to be the most suitable binding method for highly localized measurements of specific nanostructures. Both NIST and PTB have made significant efforts in dimensional nanometrology that rely on this approach. Details of this approach are also described in section 5.3.

The NIST Single Crystal Critical Dimension Reference Material (SCCDRM) project was a multi-generational effort to develop standards for linewidth measurements below 100 nm (Cresswell *et al* 2002, 2006, Dixson *et al* 2005). The purpose of this project was to establish traceable linewidth measurements for specific crystalline Si nanostructures. The measurement target is localized by a unique mutual navigation index. Only native Si oxides were present on the Si structure. A common approach was to use a critical dimensional atomic force microscope (CD-AFM) as a comparator between those structures cross-sectioned for TEM and those that remain. The expanded uncertainty was 0.6 nm (k = 2). However, the standard samples distributed to users expanded the uncertainty (k = 2) between 1.5 nm and 2 nm. The use of TEMs to achieve this overall approach and traceability has been generally accepted in the field of dimensional nanometrology, especially semiconductor metrology. In fact, vendors with secondary standards for the industry have introduced a similar approach to calibrate one of their products (Tortonese *et al* 2004).

Recently, PTB has independently implemented conceptually relevant methodologies (Dai *et al* 2013, 2015). The PTB implementation of CD-AFM is based on a method known as Vector Approach Probing (VAP) (Dai *et al* 2012). The final uncertainties in the implementation of NIST and PTB were about the same. However, given the potential for outliers, it suggests that the PTB method can potentially extend to even lower uncertainties. PTB, in collaboration with commercial standard sample vendors, has developed a standard sample for commercially available crystalline silicon linewidth standards (figure 8.5) (Dai *et al* 2015).

Accurate allocation of feature edges in high-resolution (S)TEM images remains a challenging issue, which is crucial in providing traceable criteria for dimensional nanometrology using TEM measurements. Uncertainty in the position of feature edges directly affects the uncertainty in width measurement. In general, these

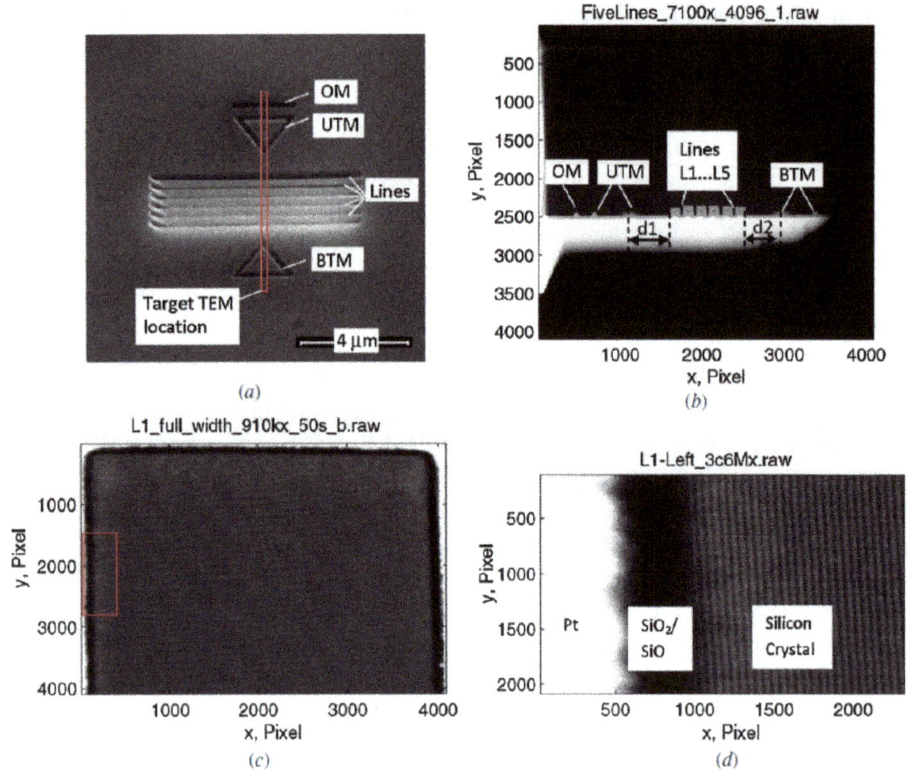

Figure 8.5. (a) SEM image showing line feature patterns and added alignment marks. (b) Overview STEM image obtained at a magnification of 7.1 k. (c) A typical TEM image measured with a CD reference structure (nominal CD 100 nm) at a magnification of 910 k. (d) A typical TEM image obtained at a high magnification of 3.6 M taken on the left side of the structure, as marked by (c) (Dai *et al* 2013). Copyright IOP Publishing Ltd. All rights reserved.

uncertainties should be less than 1 nm to maintain useful uncertainties in the final width calibration. Edge uncertainty depends on multiple factors including (1) the nature of the original sample (i.e. crystallinity), (2) potential damage, sample pretreatment performance including annealing, and (3) TEM imaging physics.

The requirement for direct traceability by decomposing the silicon lattice is that at least some parts of the sample material, ideally the main target shape, must be single crystal. This was true for both the NIST and PTB projects. However, this requirement imposes considerable limitations on sample design and production. The only comparable commercial practice used an approach where the primary target was polycrystalline, but a nearby feature that could be included in the same field of view (FOV) was a single crystal. Although it has been successfully performed, this approach relies in part on a thorough understanding of imaging performance and TEM non-linearity throughout its FOV. If the target feature is crystalline, only non-linearity over the thickness of the oxide or other capping layer can contribute to the final uncertainty. For adjacent features, the spacing is two orders of magnitude

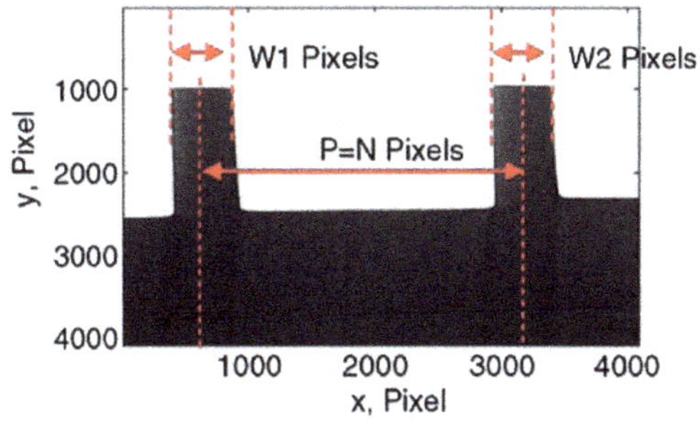

Figure 8.6. Calibration of image magnification via pitch between structural pairs. It was calibrated accurately and traceably by the metrological AFM (Dai *et al* 2013).

larger. For high-end TEM equipment, this may not seem to be very important, but it is a practical consideration for users trying to implement this approach with low-end equipment.

Another variant of the general approach also implemented by PTB is called the 'pitch method' (figure 8.6) (Dai *et al* 2013). The samples involve a pair of target structures that can be observed within the TEM FOV. Prior to performing a TEM cross-section, the pitch of these features was measured using AFM. AFM linewidth measurements are highly dependent on tip width, but pitch (one-way) measurements using AFM are primarily sensitive to scanner displacement. PTB used metrological AFMs with displacement interferometry built into the scanner to perform traceable pitch measurements to the SI meter. Since both features could be observed within the same TEM FOV, it was possible to transfer AFM traceability to the TEM scale and ultimately to the feature width. Experimental studies of PTB have demonstrated that the consistency between the pitch method and the silicon lattice method is greater than or equal to 0.3 nm (Dai *et al* 2013).

Figure 8.7 shows a strategy for non-destructive calibration of indirect line pattern features using TEM measurements (Dai *et al* 2015). The strategy consists of three steps. In the first step, two groups of samples are selected. One is selected as the reference structure and the other is selected as the TEM target structure. Two groups of samples are measured by CD-AFM using the same flare tip under the same measurement conditions. In the second step, the TEM target structures are destructively measured and their dimensions can be evaluated. In the third step, the (valid) tip shape can be evaluated based on the TEM reference value. The dimensions of the reference structure can be also calculated using the known (effective) tip shape. The reference structure has not been measured by TEM, but it can be successfully applied to reference measurements.

8.2.3 CCL-GD-MeP-3

Details of the atomic step are described in section 3.4.

Si samples in large areas with no steps and areas with single-atomic steps and wide terraces are considered to be very useful for detecting scanner misalignment and calibrating the vertical axis of equipment used for dimensional applications. If

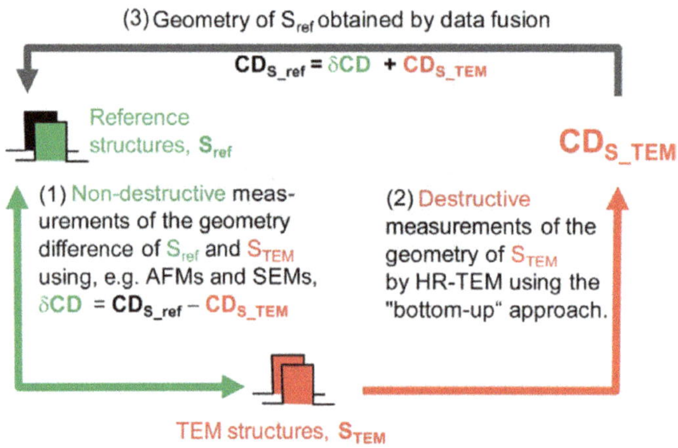

Figure 8.7. Non-destructive calibration strategy for indirect line pattern features using TEM measurements (Dai *et al* 2015). Copyright IOP Publishing Ltd. All rights reserved.

the monoatomic steps on the Si sample are prepared in a suitable manner, the vertical axis can be traceable up to SI meters through the reference values for the silicon step heights. On the other hand, the use of Si (111) and Si (100) under UHV conditions and Si (111) in air is sufficiently widespread. Further experiments on Si (100) in air are recommended. For step height values, higher uncertainty should be used instead of the very small uncertainty of bulk value. The reasons for this are related to the effects of unknown stresses and strains, and the unknown effects of oxidation. The use of improved products in the future may lead to a further reduction in the uncertainty stated. It is recommended that silicon lattice constants are used to calibrate the normal scan axes of high-resolution equipment in surface measurements. Depending on their crystallographic orientation, the following values should be used for the distance between adjacent monoatomic steps. When the surface orientation lattice parameter is d_{100}, the silicon monoatomic step height is 135 pm (expanded uncertainty ($k = 2$) 5 pm in ultra-high vacuum, uncertainty 15 pm in atmosphere), and when the surface orientation lattice parameter is d_{111}, the silicon monoatomic step height is 313 nm (uncertainty 5 pm in ultra-high vacuum, uncertainty 15 pm in atmosphere). The useful range of Si steps for calibration of surface-measuring instruments is limited to heights less than 10 nm. Further studies in the manufacturing process of monoatomic silicon step height standards may make it possible in the future to extend this calibration range and further reduce the measurement uncertainty of monoatomic Si steps.

8.3 Future directions of nano dimensional standards

The impact on the industry of showing MeP is explained using CCL-GD-MeP-1 as an example. The traceable X-ray diffractometer was introduced in section 2.3.2. The traceable X-ray diffractometer is used for pitch calibration of one-dimensional gratings for magnification calibration of scanning electron microscopes. The X-ray diffractometer shown in section 2.3.2 enables SI traceable calibration since the X-ray wavelength is calibrated using NIST SRM 2000. However, it is unclear whether NIST SRM 2000 will be available to users for a long period of time, and how to maintain traceability of X-ray wavelengths has been a long-standing issue. With the introduction of the CCL-GD-MeP-1 (secondary realization of the SI meter using silicon lattice parameter and X-ray interferometry for nanometer and sub-nanometer scale applications in dimensional nanometrology) in 2019, and the recommended value of the silicon lattice constant will be the absolute value as it is, the NIST SRM 2000 can be used as a check standard for the X-ray wavelength.

When the unstabilized He–Ne laser was registered as MeP in 2007, it became free from periodic laser frequency calibration using an I_2 stabilized He–Ne laser, which was good news in the dimensional metrological industry (Stone *et al* 2009). The three MeP guidelines have been recently recommended, and industry-friendly nano dimensional standards have been fixed. In the future, this can be expected to greatly contribute to the progress of nanotechnology and the development of calibration technology in nano dimensional standards.

8.4 Summary

Firstly, the status of nano dimensional standards was summarized. Next, the secondary realization of an SI meter using silicon lattice parameters was introduced. Finally, the impact on the industry due to the release of CCL-GD-MeP-1 was explained.

References

Basile G *et al* 2000 Combined optical and X-ray interferometer for high precision dimensional metrology *Proc. R. Soc. A* **456** 701

Cresswell M, Guthrie W, Dixson R, Allen R A, Murabito C E, Martinez and de Pinillos J V 2006 RM8111: development of a prototype linewidth standard *J. Res. Natl. Inst. Stand. Technol.* **111** 187

Cresswell M W, Bogardus E H, Martinez de Pinillos J V, Bennett M H, Allen R A, Guthrie W F, Murabito C E, am Ende B A and Linholm L W 2002 CD reference materials for sub-tenth micrometer applications *Proc. SPIE* **4689** 116

Dai G, Heidelmann M, Kübel C, Prang R, Fuegge J and Bosse H 2013 Reference nano-dimensional metrology by scanning transmission electron microscopy *Meas. Sci. Technol.* **24** 085001

Dai G, Häßler-Grohne W, Hüser D, Wolff H, Fluegge J and Bosse H 2012 New developments at Physikalisch Technische Bundesanstalt in three-dimensional atomic force microscopy with tapping and torsion atomic force microscopy mode and vector approach probing strategy *J. Micro/Nanolith. MEMS MOEMS* **11** 011004

Dai G, Zhu F, Heidelmann M, Fritz G, Bayer T, Kalt S and Fluegge J 2015 Development and characterisation of a new linewidth reference material *Meas. Sci. Technol.* **26** 115006

Dixson R, Bosse H and Yacoot A 2019 Recommendations of CCL/WG-N on: realization of SI metre using silicon lattice and transmission electron microscopy for dimensional nanometrology *CCL-GD-MeP-2*

Dixson R G, Allen R A, Guthrie W F and Cresswell M W 2005 Traceable calibration of critical-dimension atomic force microscope linewidth measurements with nanometer uncertainty *J. Vac. Sci. Technol.* B **23** 3028

Koenders A L, Busch I, Garnaes J, Yacoot A, Dixson R and Bosse H 2019 Recommendations of CCL/WG-N on: realization of SI metre using height of monoatomic steps of crystalline silicon surfaces *CCL-GD-MeP-3*

Misumi I, Sugawara K, Kizu R, Hirai A and Gonda S 2019 Extension of the range of profile surface roughness measurements using metrological atomic force microscope *Prec. Eng.* **56** 321

Mohr P J, Taylor B N and Newell D B 2012 CODATA recommended values of the fundamental physical constants: 2010 *Rev. Mod. Phys.* **84** 1527

Stone J A, Decker J E, Gill P, Juncar P, Lewis A, Rovera G D and Viliesid M 2009 Advice from the CCL on the use of unstabilized lasers as standards of wavelength: the helium–neon laser at 633 nm *Metrologia* **46** 11

Tortonese M, Lorusso G, Blanquies R, Prochazka J and Grella L 2004 Sub-50-nm isolated line and trench width artifacts for CD metrology *Proc. SPIE* **5375** 647

Watanabe H, Yamada N and Okaji M 2004 Linear thermal expansion coefficient of silicon from 293 to 1000 K *Int. J. Thermophys.* **25** 221

Yacoot A, Bosse H and Dixson R 2020 The lattice parameter of silicon: a secondary realisation of the metre *Meas. Sci. Technol.* **31** 121001

Yacoot A and Kuetgens U 2012 Sub atomic dimensional metrology: developments in the control of X-ray interferometers *Meas. Sci. Technol.* **12** 074003

Yacoot A, Kuetgens U, Koenders L and Weimann T 2001 A combined X-ray interferometer and scanning tunnelling microscope *Meas. Sci. Technol.* **12** 1660

Yacoot A, Kuetgens U, Massa E, Dixson R and Bosse H 2019 Recommendations of CCL/WG-N on: realization of the SI meter using silicon lattice parameter and X-ray interferometry for nanometer and sub-nanometer scale applications in dimensional nanometrology *CCL-GD-MeP-1*

Lightning Source UK Ltd.
Milton Keynes UK
UKHW050943150721
386952UK00006B/35

9 780750 331890